D0653589

Northamptonshire
DISCARDED
16.3.05

SPRATTON C.E.
PRIMARY SCHOOL

# Taking
# TO THE AIR

## 1900 to 1925

01774

Reader's
Digest

Published by The Reader's Digest Association Inc.
London • New York • Sydney • Montreal

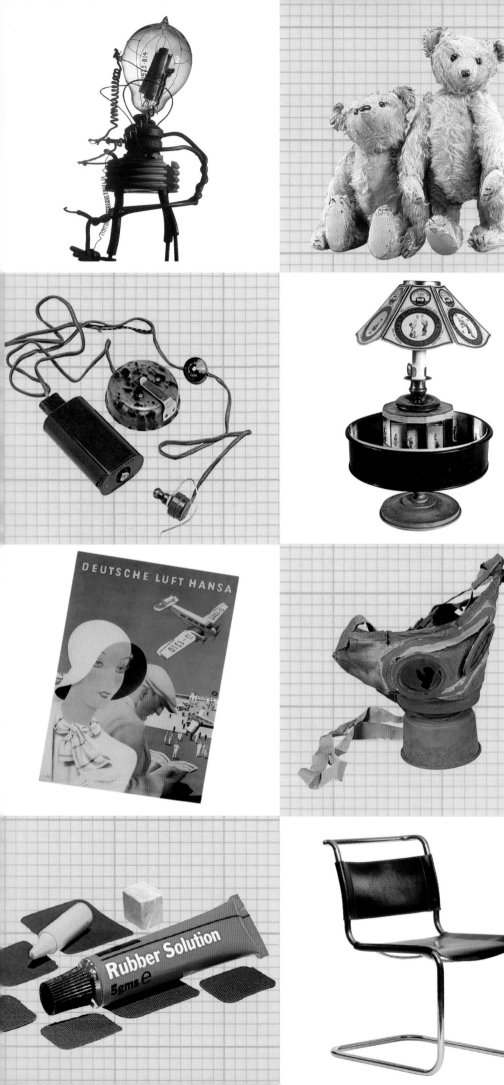

# Contents

# Introduction

As the new century dawned the Western world, which had been struggling for decades to recover from the economic depression of the 1870s, began to experience an economic growth that rekindled faith in the future. A second industrial revolution was about to begin, based on electricity and the automobile. As if in reflection of this new optimism, intrepid aviation pioneers finally realised the long-held dream of powered flight: the age of aviation had arrived.

Yet this rapidly soaring vision was set against a backdrop of some old-fashioned and reactionary forces. For this was also the age of Empire, when European colonialism was at its height. There were seething tensions within Europe itself, with volatile hotspots in the Balkans and in Russia. These soon erupted into war and revolution that would shape the 20th century. In the First World War – the first truly global conflict – science and technology were put to use in terrifying and ever more efficient weapons of mass slaughter. When the 'war to end all wars' was finally over, millions of men lay dead and three great empires of Old Europe – the German, Austro-Hungarian and Russian – had vanished in the air. The victors, economically and emotionally exhausted by the Herculean effort it had taken to win the war, punished the vanquished with crippling

**Modernism meets Vienna**
*Karlsplatz station, designed by Otto Wagner, still stands as a glorious example of the* Jugendstil, *the Art Nouveau style as expressed in Vienna. By the 1920s, Art Nouveau was giving way to Art Deco.*

demands for reparations. In Russia, the death of the old order came with the Bolshevik Revolution of 1917. Suddenly, old certainties were swept away: even the laws of physics had been turned upside-down by Einstein.

For a while, the newly formed League of Nations raised hopes of a new era of universal peace and the Jazz Age provided many with escapist relief in the form of the latest music and dance crazes. But all the frivolity and folly of the 1920s could not conceal the fact that the troubles of Europe and the world had not gone away. Meanwhile, the aeroplane – building on the urgent development spurred by the war – continued its rapid and inexorable rise.

*The editors*

KARLSPLATZ

▲ In the 1920s and 1930s radio became a mass medium as new, easy-to-use radios came on the market: this stylish Art Deco example has just three buttons – one to find the station, one to change the wavelength and one to alter the volume

▲ British scientist Augustus Waller demonstrated his prototype of the electrocardiograph to the Royal Society in London in 1887

► Teddy bears first came on the market in 1903 in both Germany and the United States and were an instant hit with children; these two examples were made by the German Steiff company in 1904

From the first tentative hops made by the Wright brothers' *Flyer* at Kitty Hawk in 1903, the aeroplane developed in leaps and bounds. Louis Blériot made the first cross-Channel flight in 1909 and just a decade later Alcock and Brown made the first flight

▲ The vials in this early example of an allergy-testing kit contained potential allergens that were dripped onto a patient's skin to test for any adverse hypersensitive reaction

▼ The clementine was produced by crossing a bitter orange with a mandarin and is named after the man who first cultivated it, Fr Clement

▼ The speedometer, invented in 1902, quickly became standard equipment in cars, as featured in this elegant walnut dashboard of a Bugatti touring car of the 1920s

across the Atlantic Ocean. That same year, 1919, saw the first commercial flights offering seats to paying passengers. The aeroplane was shaping up into the most remarkable technological leap forward of the era. The earliest aviation pioneers entrusted

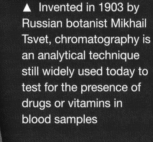

▲ Invented in 1903 by Russian botanist Mikhail Tsvet, chromatography is an analytical technique still widely used today to test for the presence of drugs or vitamins in blood samples

SÉDOSINE
nervosité - angoisses - insomnies

▲ Barbiturates came on the market as sleeping aids from 1903 and were widely used by psychiatrists in narcotherapy

▲ Once the role of hormones had been identified, in 1902, chemists began synthesising them for medical purposes: the insulin preparation Iletin was introduced in 1923 to help diabetics control blood sugar

their lives to flimsy machines made of canvas and bamboo, but with the development of light, powerful rotary engines and reliable navigational aids, flying gradually became a less risky business. Another major breakthrough was wireless technology, which soon

▲ The technique of offset litho printing, discovered by chance in 1904, was cheap yet highly effective; this handbook for printers and typographers was produced by the Bauhaus in Germany in 1926

▲ A modern re-enactment of the pioneering escapades of Wilbur and Orville Wright, who in 1903 succeeded in making the first controlled flight in a heavier-than air machine – the *Flyer*

▶ The city of New Orleans was the birthplace both of jazz and of the jazz trumpeter Louis Armstrong, seen here in his Chicago days in the 'Hot Five'

replaced telegraph cables and telephone lines in many key roles. Radio was at first restricted to ship-to-shore communications, but before long it became a medium of mass entertainment and information. In this era of radical upheaval, traditional conceptions

► Mass vaccination programmes against tuberculosis were undertaken in the 1920s, following development of the BCG vaccine in 1906

INDICATOR

▲ The prototype of sonar, which was used extensively by shipping in the Second World War (above), was developed in 1905 by a Norwegian, Einar Berggraf

of time and space were turned on their head by Albert Einstein and his theory of relativity, which demonstrated that the passage of time was not, as previously assumed, a universal and constant phenomenon. Many of the period's technological advances found

◄ The technique of freeze-drying, invented in 1906, came to be used for a wide variety of food products including soup

▼ Reinforced concrete, patented in 1892 by Frenchman François Hennebique, transformed construction in the 20th century, enabling architects to build far more ambitious structures like this enormous hangar for airships built at Orly in Paris (now demolished)

◄ Albert Einstein revolutionised the way that scientists think of time and space and in the process laid the foundations of quantum physics

◄ Mickey Mouse made his debut in 1928 and launched his creator Walt Disney to the forefront of animated cartoon films; before Mickey the great star of the silent cartoon era was Felix the Cat, who first appeared in 1919

military applications during the First World War. The conflict boosted the development of aviation and made extensive use of telecommunications, but it also spawned some terrifying new weapons – notably the tank and poison gas. After the war,

► Half-boat, half-plane, the hydrofoil – seen here on its first test on Lake Maggiore in Italy in 1906 – soon smashed the world speed record on water and the principle was later taken up for passenger craft, naval patrol vessels and racing yachts

◄ Workers at Henry Ford's Highland Park factory in Michigan assembling parts for the Model T; by applying 'scientific management' to operations on the factory floor, Ford began to push down the price of the car that would bring motoring to ordinary people

► A telegraphoscope receiver of 1907, a precursor of the fax machine; a refined and more compact version of this device, the Belinograph, became the ancestor of today's fax machines and photocopiers

people's quality of life slowly began to improve once more. In the United States Henry Ford was revolutionising car manufacturing to bring motoring to the masses in the form of the Model T, the first affordable car; following the introduction of the assembly line, the

► The age of plastics dawned with the invention of Bakelite in 1909; the fact that the new material could be moulded into virtually any shape made it a favourite with designers and it soon found its way into desk lamps, toys, telephones and much else

MICKEY MOUSE
PRESENTS
A WALT DISNEY
SILLY SYMPHONY
FLOWERS AND TREES
UNITED ARTISTS PICTURE

◄ The cartoon *Flowers and Trees*, directed by Burt Gillett for the Walt Disney corporation in 1932, was the first film made using the three-strip Technicolor process

▲ In 1910 zoologist Thomas Hunt Morgan showed that physical traits, such as the colour of our skin, hair and eyes, are passed on through genes carried on sex chromosomes

cars rolled off the production lines by the thousand. More cars needed more roads, reflected in the arrival of the first motorway in Germany. Medical knowledge advanced apace in this era, with the discovery of allergies, hormones, genes, vitamins and barbiturates.

▼ The value of fresh fruit came to be better understood after the first vitamin – thiamine – was identified by biochemist Casimir Funk in 1911

▲ According to the theory of continental drift, first described by German meteorologist Alfred Wegener in 1915, this is how the continents were placed some 65 million years ago

**AN APPLE A DAY KEEPS THE DOCTOR AWAY**

◄ A quintessential symbol of the Jazz Age, neon signs began to appear in cities from the 1920s

► The first tank unleashed in combat was the British Mark I at the Battle of the Somme in 1916, but it was the Renault FT-17 (right) that established the basic pattern for future tank design

Meanwhile, the world of entertainment was being transformed by the rise of jazz – born in New Orleans – and animated cartoons. In the domestic sphere, as women returned from their war work, a raft of new inventions awaited them intended to ease their role as

▼ The intrepid pilots of early airmail services undertook dangerous flights around the world, including crossing huge mountain ranges such as the Andes in South America

◄ Gas masks – this one is a French example from 1918 – offered little protection to soldiers against mustard gas, which the Germans used for the first time in warfare in 1917 during the Second Battle of Ypres

homemakers. Convenience was becoming the quest of many innovators, as freeze-dried foods, self-service shops and easy-care materials like stainless steel and plastics all emerged. In the cities, the dawning of the age of modernity announced itself in

► Early aviation pioneers took incredible risks, like the German glider pilot Otto Lilienthal, who in 1893 became the first person to be photographed in flight (right), but died three years later following a gliding accident

▼ Sigmund Freud's famous consulting couch; he sat alongside in the green armchair as he practised the therapeutic technique known as psychoanalysis

▲ An electric toaster from the 1920s

soaring reinforced concrete buildings and garish neon lighting. It was not just the physical world that was changing: since the turn of the century, Sigmund Freud had been exploring the unconscious mind and developing his therapeutic technique of psychoanalysis.

▲ While Freud was exploring the unconscious mind, the German psychiatrist Hans Berger began measuring brain activity by means of his invention, the electroencephalogram (EEG)

▲ The first self-service supermarkets opened their doors in the United States in 1910; designed to maximise profits, they also proved popular with customers by making shopping less time-consuming

▶ Vienna in the early 20th century was a fascinating mix of tradition and modernism and nowhere was this more evident than in its architecture, as Baroque splendour rubbed shoulders with radical new buildings; Otto Wagner's Karlsplatz station (right) expressed the Viennese take on Art Nouveau

Freud was a resident of Vienna, the Austro-Hungarian imperial capital, left without an empire after the First World War. It was a city that encapsulated the dynamic yet febrile spirit of age, where traditional forces mingled uneasily with the European avant-garde.

# THE STORY OF INVENTIONS

The two most momentous inventions of the early 20th century were radio and the aeroplane. The first wireless sets were unwieldy objects, whose signal could barely be heard above a constant crackle of static. And yet within three decades radios could be found in virtually every home in the developed world. Likewise, the flimsy contraptions that took the early pioneers of flight into the air were soon to change for ever the way people travelled. These two technologies were developed at the same time, marking the onset of a new era – the age of communication.

# Opening up the airwaves

Guglielmo Marconi, the father of wireless telegraphy, established the world's first radio link in 1899. The first person to transmit speech by radio was the Canadian physicist Reginald Aubrey Fessenden, thanks to his discovery of the modulated carrier wave.

**Primitive detector**
*This device, known as Branly's 'coherer', was used by its inventor to detect radio waves produced by a spark generator some 20 metres away.*

**Backroom boffin**
*Édouard Branly in 1900 in his Paris laboratory, which was later turned into a museum.*

'One, two, three, four. Is it snowing where you are Mr Thiessen? If so telegraph back and let me know.' Thus ran the message sent by Reginald Fessenden to his assistant on 23 December, 1900. Fessenden was in his laboratory at the meteorological station on Cobb Island in the Potomac River, speaking into a microphone connected to a radio transmitter he had designed and built himself. More than half a mile away, Thiessen listened to this first ever transmission of intelligible speech by electromagnetic waves and replied, in Morse code, that it was indeed snowing.

## Cumulative effort

Fessenden's achievement built on the work of several predecessors. The electric telegraph had blazed a trail but, like the telephone that followed it, its development was hampered by the high cost and huge difficulties involved

---

### A PROLIFIC INVENTOR

Reginald Aubrey Fessenden, seen here (right) in a photograph taken c1910, was born in Quebec in 1866. From an early age, he showed a gift for science and studied to become a physicist. In 1886 he moved to New York in the hope of securing a position with the inventor Thomas Alva Edison. Although reluctant at first, Edison eventually hired him to work in his new laboratory in West Orange, New Jersey. Four years later, Edison found himself in financial difficulties and let several employees go, including Fessenden. After a spell as a university lecturer, Fessenden joined the US Weather Bureau in 1900, where he

planned a chain of coastal wireless stations. Despite lasting only two years, this post saw him achieve many of his major breakthroughs in radio. Thereafter, with private backing, he founded the National Electric Signaling Company (NESCO), but the enterprise failed to develop any commercially successful spin-offs and closed in 1911. This was Fessenden's last foray into radio and he now threw his energies into other inventions, filing more than 500 patents. These included an underwater oscillator – a precursor of SONAR – that was used in anti-submarine warfare during the First World War. He died in 1932.

in laying cables. Inventors began searching for ways round this problem, while physicists posed fundamental questions about the nature of electricity. Their work bore fruit in the form of the wireless telegraph.

In 1887 the German physicist Heinrich Hertz had discovered that light waves were not the only form of electromagnetic waves. His findings were confirmed three years later by Frenchman Édouard Branly and his invention, the 'coherer'. This rudimentary radio detector consisted of a small glass tube filled with iron filings; under the influence of electromagnetic radiation, the resistance of the filings reduced sharply. In 1894 the Russian Alexander Popov built his own version of the coherer with a vertical antenna – the world's first true radio receiver.

## Marconi and wireless telegraphy

Despite the best efforts of trained scientists, the link between Hertzian radiation (radio waves) and the telegraph would be made by a self-taught inventor with no scientific education. Guglielmo Marconi was born in 1874 near Bologna and began experimenting with radio waves in 1894. He was responsible for two major innovations – adopting Morse code from telegraphy and using the antenna for both transmission and reception. Within a year, Marconi had succeeded in sending signals over long distances, demonstrating the importance of the height of the antenna. His aim was to use radio waves to transmit messages to ships at sea.

The Italian authorities showed little interest in Marconi's work, so he went to England. From there, in 1899, he radioed a greeting across the Channel to Édouard Branly. Both the British army and Royal Navy were impressed and immediately took a stake in his invention. In 1901 he made the first transatlantic radio transmission, from Poldhu in Cornwall to St John's in Newfoundland. Before long other scientists and engineers, including Popov in Russia, Adolf Slaby in Germany and Eugène Ducretet in France, devised similar systems for use on board naval vessels. Yet wireless telegraphy still had serious limitations: at best, it could transmit only 15 words per minute (compared to 500–600 by telegraph) and the quality of reception was very poor.

## Transmitting the human voice

To achieve his main aim of transmitting the human voice audibly and clearly, Fessenden relied upon a principle that had only recently been established.

Marconi did not have a firm grasp of wave theory and so thought of the brief signals emitted by his spark-gap transmitter as being like invisible explosions. His apparatus worked fine for sending the dots and dashes of Morse code, but not for reproducing the modulations of the human voice or of music. Fessenden therefore conceived the idea of transmitting a continuous signal, or sinusoidal wave, whose modulations would encode those of the human voice. Here, then, was the concept of the carrier wave, which has been used for radio transmission

**Father of radiotelegraphy**
*Marconi in the radio room of his steam-yacht* Elettra. *This vessel, which served as a floating laboratory for the Italian inventor, was requisitioned by German forces in Italy in 1943.*

**Great leap forward**
*The thermionic valve was invented in 1904 by the English physicist and engineer John Fleming when he was scientific adviser to the Marconi Company. The device contributed enormously to wireless technology (above left), and was only replaced some 50 years later by the transistor.*

21

## DIODES AND OTHER VALVES

The diode is a valve fitted with two electrodes, which is capable of detecting radio waves. The triode valve developed by Lee De Forest had, as its name implies, a third electrode. This made it more sensitive and also enabled it to be used as a transmitter. The tetrode (with four electrodes) was invented during the First World War by Walter Schottky of the Siemens and Halske Company in Germany and was used for receiving high-frequency signals. In 1926 Bernard Tellegen of Philips introduced the pentode (five electrodes), which solved the problems of overheating to which the tetrode was prone.

### Major advance
*Lee De Forest holding a triode, a device that greatly amplified the radio signal.*

**Crude but effective**
*An experimental lamp of 1889 (right) used by John Fleming in his research into electrical conduction. Fifteen years later, those experiments would inspire him in creating the first thermionic valve, also known as the 'Fleming valve' or diode.*

ever since. Fessenden's apparatus consisted of high-frequency transmitters made from modified alternators fitted with antennae. By placing a microphone between the generator and the antenna, he found he could modulate the amplitude of the carrier wave.

On Christmas Eve in 1906, Fessenden achieved a major breakthrough when he used his high-frequency alternator to transmit speech and music (a passage from the Bible, recordings of Handel and his own rendition of *O Holy Night* on the violin) to a dozen or so ships of the United Fruit Company, which he had equipped with receivers. His transmission, made from Brant Rock in Massachusetts, was the world's first radio broadcast.

### Towards the transistor

Improvements were made in radio technology little by little. In 1904, for example, the English electrical engineer John Ambrose Fleming invented the two-electrode vacuum tube, a rectifier that made current flow in a single direction. The 'Fleming valve', or diode,

## TRANSMISSION FREQUENCIES

Marconi soon came up against the problem of interference on wireless transmissions. John Fleming realised that the problem was caused by all spark gaps transmitting at the same frequency and proposed that each transmitter should henceforth have its own unique frequency, which the receiver could pick up by tuning in to the same wavelength. On the eve of the First World War, all the combatant nations set aside particular frequencies for military communication, so ushering in the age of electronic espionage, encryption and countermeasures.

supplanted Branly's 'coherer'. Three years later, the Americans Lee De Forest and Edwin Armstrong unveiled the triode, or 'audion'. This amplifier vacuum tube was based on Fleming's invention but had a third element, the grid, between the anode and cathode. The grid allowed modulation of the current through the valve with minimal voltage changes.

The earliest wireless sets consisted of just an antenna, a lead sulphide (galena) crystal to detect the radio waves, and headphones. They were cheap and self-powered, but could only be listened to by one person and reception was weak. From 1913 they were replaced by valve radios incorporating triodes, loop antennae, loudspeakers and batteries (audions used a lot of power). These were more substantial devices that gave far better reception. Housed in elegant wooden cabinets, they made luxury items for middle-class homes.

The discovery of shortwave high-frequency (HF) radio waves in 1923 and of ultra high-frequency (UHF) in 1930 opened up new horizons. By reflecting these waves off the

ionosphere – the upper layer of the Earth's atmosphere – it was found that a signal could be picked up far from the transmitter. The development of cheaper 'midget' radios in the 1920s and 1930s, which sold in their tens of millions, greatly boosted wireless ownership.

**Cat's whisker**
*A crystal radio set from the early 20th century (left). Crude radio receivers like this were easy to assemble by amateurs and were widely used by resistance groups in occupied countries during the Second World War.*

**Personal radio**
*In the 1920s, light, portable radio sets with earpieces became an indispensable fashion accessory.*

# Turn on, tune in

The soundtrack of the 20th century, a time of great change and turmoil, was provided by radio, the very first mass medium. Radio rapidly became an indispensable part of people's lives as they gathered round the wireless to enjoy dance music or dramas, or to listen anxiously for news of war.

**The airwaves at your fingertips**
*A 1925 poster for the 'Radiola' set manufactured by the Société Française Radioéléctrique (SFER). With just the simple turn of the Radiola's dial, the listener can tune in to stations across Europe.*

UNE SEULE MANŒUVRE avec le "SFER-20" RADIOLA

**Save Our Souls**
*In this reconstruction for the camera, telegrapher David Sarnoff receives a distress call from the sinking Titanic. The emergency signal was broadcast in Morse code from the stricken ship both as 'SOS', which had been adopted as the universal distress signal in 1908, and as 'CQD', a distress signal introduced by Marconi in 1904. SOS was adopted for its ease of recognition; its English acronyms were devised later.*

Dot-dot-dot … dash-dash-dash … dot-dot-dot, the famous Morse code distress signal, was first introduced in Germany in 1905. It was adopted as the international distress call in 1908, when it quickly became known as 'SOS'. Henceforth, it would play a major part in saving people's lives around the world.

On the night of 14–15 April, 1912, the American telegrapher David Sarnoff picked up the distress call from the ill-fated ocean liner RMS *Titanic* after she struck an iceberg off Newfoundland. Sarnoff alerted other ships in the area, which managed to pick up 865 survivors. Wireless telegraphy assumed a vital role in the First World War, providing communication links for warships, submarines and aircraft, as well as for infantry in the trenches. As the technology improved, the wireless telegraph was superseded by radio. In January 1918, US President Woodrow Wilson used a radio broadcast to unveil his Fourteen Point Peace Plan to the peoples of Europe.

## Early public broadcasts

After the First World War radio became big business. The European market was carved up

## INDEPENDENT VERSUS STATE RADIO

Private radio stations mushroomed in the USA in the 1920s and 1930s. Broadcasters adopted an informal chatty style and the stations played a key role in the rise of jazz. By contrast, in Europe state monopolies in radio made for a more staid, elitist form of broadcasting. After the Second World War, as more independent stations came on air in Europe, radio became more entertaining and less didactic.

**Popular propaganda**
*Adolf Hitler, Joseph Goebbels (seated) and Gauleiter Adolf Wagner (left) listen in to a radio set on 13 January, 1935, as the result of the Saarland plebiscite is announced. This coalmining region on the border of France and Germany had been occupied by France since the end of World War I. The plebiscite showed more than 90 per cent of its people to be in favour of rejoining the German Reich.*

## ROOSEVELT'S 'FIRESIDE CHATS'

**A**fter the 1929 Wall Street Crash, the world sank into a severe economic depression. In the USA, President Franklin D Roosevelt launched the 'New Deal' to get Americans back to work. To publicise this programme, he asked the country's major broadcasting networks to air a series of so-called 'fireside chats' in which he addressed the nation on a personal level. Speaking calmly from his fireside chair, Roosevelt struck just the right note, delivering his political message with folksy snippets from daily life.

between large concerns such as the Marconi Wireless Telegraph Company in Britain, Telefunken in Germany and CSF in France. In the USA, the giant Radio Corporation of America (RCA) was founded in 1919.

The radio bug bit America straightaway, with amateur radio enthusiasts using war-surplus equipment to send out messages, music and local news. Spotting a niche in the market,

RCA commissioned one of those radio hams, Frank Conrad, to make regular broadcasts. His transmitter, with the call sign KDKA, became the world's first commercial radio station. On 2 November, 1920, it announced the election of President Warren Harding.

David Sarnoff, by this stage a senior executive at RCA, dreamt of equipping every home in America with a radio set. As a way of promoting the radio, he organised several high-profile radio transmissions, most notably the world heavyweight title fight between Jack Dempsey and Georges Carpentier in July 1921. The sales of wirelesses rocketed.

The first European radio stations came into existence soon after. Radiola, a private French station that later became Radio Paris, began broadcasting on 6 November, 1922. Eight days later, on 14 November, the BBC went on air in Britain with its first daily service, 2LO.

### Inform, entertain … and coerce

The very first transmissions were musical or educational, but before long radio stations began broadcasting news. The first BBC news broadcast was made by Arthur Burrows, director of programmes. Each bulletin was read twice – once quickly and once slowly – and listeners were asked to say which they

preferred. There was an eclectic mix of content: a speech by the Conservative leader Andrew Bonar Law, details of Old Bailey sessions, a report of a train robbery, the sale of a Shakespeare folio, fog in London – and the latest billiards scores.

Growing international tensions in the 1930s contributed to the steady rise of radio as a news medium. On 12 March, 1938, as German troops marched into Austria, a French radio reporter broadcast live from a telephone in Vienna, with the crunch of marching jackboots clearly audible in the background. American radio networks lost no time in sending their own correspondents to Europe.

Meanwhile, totalitarian régimes exploited radio for their own ends. Beginning in 1922, radio in Russia was an instrument for 'educating the masses'. Since very few Russians owned radios, people gathered *en masse* to listen to public broadcasts over Tannoy systems. In Germany the Nazis took over the country's radio soon after coming to power in January 1933. Tannoy systems on the streets relayed a diet of propaganda – speeches by leading Nazis, broadcasts of party rallies and endless martial music. In 1935 the *Volksempfänger* ('people's receiver') was launched; this cheap, mass-produced radio set was designed to receive only those stations sanctioned by the German state. The militaristic regime in Imperial Japan and Fascist Italy adopted a similar approach.

### FREQUENCY MODULATION

**D**eveloped by Edwin Armstrong in 1935–36, FM involves varying the frequency of the carrier radio wave while keeping its amplitude constant. This makes the signal less susceptible to disturbance.

**Object of desire**
*In the 1930s radios became highly desirable and decorative consumer goods. The valves and loudspeaker were housed in Art Deco style wooden cases made of walnut, cherry or maple.*

**Rallying call**
*From his exile in Britain, Charles de Gaulle broadcasts to his countrymen across the Channel on 30 October, 1941. De Gaulle delivered his daily 5-minute address, 'Honneur et Patrie', from BBC studios in London.*

Home front
*In a war-time scene familiar in every British home, a family gathers to listen to a speech by Winston Churchill. This particular broadcast was made on 19 May, 1945, by which time the war in Europe was over and people wanted to know when their loved ones would be coming home.*

## WAR OF THE WORLDS

On 30 October, 1938, a CBS radio correspondent reported a meteorite landing at Grover Mills, New Jersey. He described looking into the crater and seeing a silver cylinder open and tentacles emerge. The Martians had landed and were attacking – live on air. In fact, the broadcast was an adaptation of *The War of the Worlds,* the classic 1898 science-fiction novel by H G Wells. The director Orson Welles aimed for maximum realism, which included the reporter apparently being sick at the sights he was witnessing. The microphone fell from the reporter's hand as he was killed, followed by dead air. Unfortunately, thousands of people who tuned in once the drama was underway did not realise that what they were listening to was fiction. Widespread panic ensured, as people tried to flee New York. Unwittingly, Welles had demonstrated the huge power that radio could exert over its audience.

## War of the airwaves

As soon as the Second World War broke out, governments of the combatant nations assumed control over radio broadcasting. After Germany overran France in early summer 1940, the French capitulation was announced over the radio by Marshal Pétain on 17 June. On the very next day, speaking from London, the Free French leader General Charles de Gaulle broadcast his famous appeal on BBC radio for the French nation to resist.

As Britain itself came under threat of Nazi invasion, Churchill made radio broadcasts to boost the nation's morale; many of the rousing speeches had originally been delivered in the House of Commons. During the Battle of Britain and the Blitz, the European correspondent of CBS radio, Edward R Murrow, sent regular bulletins to America from London. Murrow's vivid accounts of the courage and resilience of the British people under fire helped to swing American public opinion against Nazi Germany. As the conflict

**Radio spectrum**
*The tuning dial of a radio manufactured by the German firm Grundig in 1979 shows the four different types of wavelength that the radio could receive: long-wave, medium-wave, short-wave and ultra-high frequency (L, M, K and U).*

27

## RADIO HAMS

**A**mateur broadcasters, popularly known as 'radio hams', were a feature of the very earliest days of radio in the United States. Thousands of enthusiasts got hold of crystal radio sets, built transmitters and began sending messages. Long before the advent of computers and the Internet, this was a prototype form of 'social networking'. Hobbyists were able to communicate with friends made over the air all around the globe. There are currently estimated to be some 3 million radio hams active worldwide.

dragged on, the BBC became a vital news lifeline for people not just in Britain but in occupied Europe. The station used expatriate journalists and writers to broadcast in every European language. In response, the German authorities tried jamming the signal. They also confiscated radio sets and made listening to the BBC an offence punishable by imprisonment or even death. Despite such measures, large numbers of people defied the ban. Coded messages were broadcast to let resistance forces know of planned supply drops or of rendezvous with British secret agents. The most famous of these enigmatic messages, sent on 1 June, 1944, was a quotation from the poet Paul Verlaine: 'The long sobs of autumn's violins wound my heart with a monotonous langour.' This was the signal to the French Resistance that the D-Day landings were about to take place in Normandy.

## Radio conquers the world

Frequency modulation was introduced in the 1930s and radio sets capable of receiving FM came onto the market in 1940. Transistor radios first appeared in the late 1940s. Because all their components were standardised, transistor radios were cheap to produce. Their compact size and battery power enabled people to have a radio with them wherever they went. By the early 1960s, most homes in the West had a transistor. Around the same time, the first African and Asian radio stations came on air.

Radio was a key player in the Cold War. The Voice of America, Radio Free Europe and later Radio Liberty broadcast a staunchly

**Different radios for different times**
*A compact Bakelite transistor radio (above), made by Murphy in the 1950s, came in a leather case with strap for carrying around. In contrast, African-American youths in the 1980s took to carrying huge radiocassette players pumping out bass-heavy rap and reggae (top). The machines soon acquired the nickname 'ghetto-blasters'.*

**Free speech**
*Some 'pirate' radio stations were founded to campaign for political or social change. The free station at Longwy in Lorraine, France, was set up by French unions to protest against closure of the steelworks there in the late 1970s.*

anticommunist message. The Soviet Bloc responded by jamming signals and setting up their own networks. Radio Moscow took a pacifist stance, calling for nuclear disarmament, decolonisation and global revolution. Despite the spread of television, radio remained the main ideological battleground for the superpowers. The 1960s and early 1970s saw a huge growth in 'pirate' radio stations in Europe, such as Britain's Radio Caroline, which championed rock music and opposed the state stranglehold over the airwaves.

# Clementines 1900

The clementine takes its name from a French priest, Father Clement (1829–1904). Before taking holy orders, Vincent Rodier (as he was then known) was head gardener at an orphanage at Misserghin, near the port of Oran in the French colony of Algeria. Citrus trees flourished in the balmy climate of North Africa and Rodier's orchards were laden with mandarins. The star of the show was one particular tree, around 5 metres in height, which every season bore an impressive crop of unusually small fruit with bright orange skins. The fruit was easy to pick and peel, and the flesh was far sweeter than that of other citruses; also, unlike mandarins, they had no pips. The tree was a hybrid, the result of natural cross-pollination between a mandarin and a bitter orange.

By the late 20th century, clementines had become far more popular than mandarins or tangerines. Growers in Spain, Italy, North Africa, Israel and Corsica have turned large acreages of orchard over to them. Clementines typically begin to arrive in the shops in northern Europe in November and are a favourite fruit around Christmas.

**New variety**
*The natural hybrid clementine (left) has itself been used to create an artificial hybrid, the clemenvilla, a cross between a clementine and tangerine. Clemenvillas have a thin, easy-peel skin and juicy, pip-free flesh.*

### HYBRID FRUIT

Artificial cross-pollination has enabled fruit growers to offer an ever wider range of tempting citrus fruits. The pomelo, for example, which was created on the Dutch Antilles in the late 18th century, is the result of crossing a grapefruit with an orange. The citron, meanwhile, is a hybrid of the mandarin and the pomelo.

# The Cinéorama 1900

Invented for the 1900 World's Fair in Paris by Raoul Grimoin-Sanson, the Cinéorama was the world's first panoramic film projection system. A ring of white screens 100 metres in circumference was set up beneath the Eiffel Tower. To simulate a hot-air balloon ride, the 200 spectators stood on a viewing dais that mimicked a large balloon basket. To aid the illusion, above them rose the lower half of a large gas-bag. Ten synchronised movie projectors then projected onto the screens film of the city from the perspective of an ascending balloon. The Parisian authorities closed the exhibit after just four days because of a serious fire risk, bankrupting its creator.

**Failed venture**
*Contemporary engraving of the Cinéorama by Louis Poyet in the journal* La Nature.

# Teddy bears 1903

**Marque of success**
*Vintage Steiff bears, such as these examples from 1904 (right), command high prices at toy auctions. The Steiff company, which is still in business today, put a small metal button in the ear of every bear to mark it as a genuine Steiff teddy.*

The claim to be the home of the teddy bear has long been disputed between the USA and Germany. In 1902 American president Theodore Roosevelt, while on a hunting trip in Mississippi, ordered the mercy killing of a wounded black bear. Aided by cartoons of the incident, the story fired the public imagination and inspired two toy manufacturers, Rose and Morris Mitchom, to market the first 'Teddy bear' in 1903. At the same time, a prototype of a stuffed toy bear, called 'Bär 55B', was exhibited in Germany at the Leipzig Toy Fair. Its creator, Margarete Steiff, had previously produced a stuffed toy elephant

### GUMMI-BEARS

Hans Riegel founded the Haribo sweet company in 1920. Two years later he created the 'Dancing Bear', a fruit gum in the shape of a bear. The product, which later became known as the Gold Bear or Gummi-bear, was a huge hit and is still sold the world over.

in 1880. The first Steiff bear thus also has a fair claim to be the world's first teddy bear.

Teddy bears now come in all shapes and sizes. Bears from the 1920s were typically covered with mohair, stuffed with straw or kapok, with boot buttons for eyes, long arms, a hump on their back, and often movable joints. In the 1960s the hump disappeared, the belly grew more prominent, the fur became synthetic and the stuffing mostly foam; the eyes began to be made from glass or plastic. What has not changed is the popularity of the teddy bear, which has now been a fixture of childhood for well over a century.

**Hearing in style**
*The earpiece and amplifier of a 1929 hearing aid, moulded from celluloid made to look like tortoiseshell.*

# Hearing aids 1902

Since the 13th century, people who were hard of hearing had used hollowed-out animal horns to help them hear better. By the 17th century, horns had been supplanted by specially made ear trumpets. Thomas Edison's invention of the microphone, which could record and reproduce soundwaves, in 1878 would make possible a revolution in artificial hearing. In 1902 Miller Reese Hutchinson, an associate of Edison, made the world's first electronic hearing aid, the Acousticon, comprising a microphone,

amplifier, earpieces and battery. The drawback of Hutchinson's prototype was that it weighed 12 kilos and had to be set on a table to work properly. More portable hearing aids had to await the invention of the transistor in 1947. Several more years elapsed before components were miniaturised into a device small enough to sit directly behind the ear. The first cochlear implant came in 1978; this surgical procedure involves inserting an amplifier directly into the inner ear, linked to an external microphone hidden behind the ear. This is commonly used to help people with total or partial deafness regain at least some of their hearing.

# The electrocardiograph 1902

Ever since Luigi Galvani's pioneering studies in the late 1700s, scientists knew that muscle contractions generated weak electrical impulses. The Scottish electrical engineer Alexander Muirhead was the first person to attempt to measure the electrical activity of the heart when, in 1872, he attached wires to the wrists of a feverish patient at St Bartholomew's Hospital in London. Around the same time the physiologist John Burdon Sanderson recorded results from a Lippmann capillary electrometer, a device with tubes filled with mercury able to measure small surges of electrical current.

Following these experiments, a systematic approach to producing electrocardiograms was made by British neurophysiologist Augustus Waller, who not only placed electrodes on the patient's skin but also attached the Lippmann capillary electrometer to a projector to produce a trace of the electrical activity. After trying the method on his dog, Waller applied it to create the first human electrocardiogram (or ECG, a term coined later by Dutch physiologist Willem Einthoven to describe the trace left by the heartbeat on a capillary electrometer).

## The string galvanometer

Einthoven recorded the first true electrocardiograph in 1902. To enhance the electrical impulses, he used a galvanometer, originally invented by Galvani but perfected by Clément

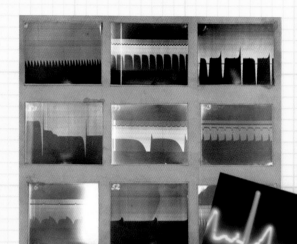

**Heartbeat detector**
*Photographic plates of heartbeats – some from patients undergoing open-heart surgery – made by Augustus Waller in the course of his researches at St Mary's Medical School in London.*

Ader. Einthoven's string galvanometer incorporated a long, silver-coated quartz filament between two powerful horseshoe-shaped electromagnets. As the current passed through, the movement in the filament was projected onto a thin photographic plate, indicating the differences in electrical potential generated by the heart's contractions. Einthoven also worked out exactly where the electrodes should be attached to the skin in order to obtain the most accurate reading. Having identified, in 1895, five distinct types of wave emanating from the heart, he devoted the rest of his life to studying and interpreting anomalies in heart rhythm.

Einthoven was awarded the Nobel prize for physiology and medicine in 1924. The device he invented has become a vital tool in the prevention and monitoring of a wide range of cardiac diseases, including myocardial infarctions, angina and heart abnormalities.

**Lifeline**
*An ECG trace showing the start of a cardiac arrest, which appears on the screen monitor as an unbroken flat line. Electrocardiograph machines first went into production in 1908.*

**His master's machine**
*Augustus Waller demonstrating his capillary electrometer to the Royal Society in London in 1887. To the enormous relief of the assembled luminaries, his dog Jimmy, who had one front and one back paw immersed in saline solution containing the electrodes, was not electrocuted.*

## PRECISION WAVES

Electrocardiographs measure the depolarisation (contraction) and repolarisation (relaxation) of the heart's auricles and ventricles. The first machines, which required water cooling aparatus for the electromagnets, weighed in excess of 270 kilograms. Over time, ECG readings increased in precision as components were miniaturised.

# Identifying allergens

In the early 1900s, scientists found that such diverse ailments as hay fever, intolerance to certain foods and adverse reactions to vaccines all had the same root cause – allergies.

**Testing kit**
*Up until 1930, a set of vials like this (right) was the standard test for allergies. Each vial contained a potential allergen; the doctor placed a drop on the patient's skin then looked for any inflammation or redness.*

**Dangerous organism**
*The tentacles of the Portuguese Man o'War* (Physalia physalis) *can be up to 30 metres long and are equipped with stinging cells known as nematocysts (visible here as small yellow spots). The sting is extremely painful to humans and, in rare cases, can be fatal.*

As the 20th century dawned, medical researchers, inspired by the earlier work of Edward Jenner on smallpox and Louis Pasteur on rabies and anthrax, began to investigate infectious diseases and their prevention through the use of vaccines.

## The discovery of anaphylaxis

In 1901 Prince Albert I of Monaco invited Charles Richet, a professor of physiology, and zoologist Paul Portier to join him on a voyage of scientific exploration on board his yacht *Hirondelle II*. Off the Cape Verde islands, the prince drew their attention to the toxin produced in the tentacles of the Portuguese Man o'War, a jellyfish-like marine invertebrate with a sting that could be fatal. Richet and Portier collected specimens with a view to creating a vaccine, but their experiment failed: the dogs that they inoculated with small doses of the poison – in theory, not enough to kill them – all died. The scientists concluded that the hypersensitive reaction of the dogs was the result of too great an interval between injections. Surmising that other toxins could produce a similar effect, Richet dubbed the severe adverse reaction 'anaphylaxis', a term from Greek meaning 'against protection'.

Richet presented the findings to the French Biological Society in May 1902, but met with a total lack of interest. Several years later, the importance of the discovery was recognised

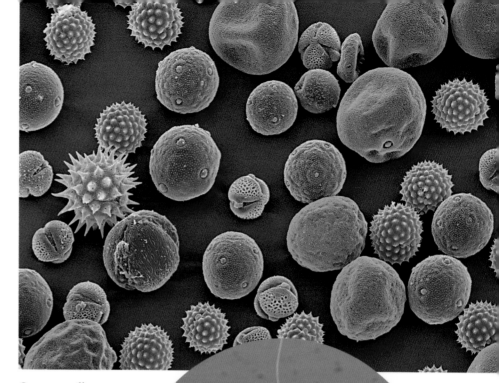

and Richet was awarded the Nobel prize for medicine for his efforts. Anaphylactic shock is the most severe form of allergic reaction known and constitutes a medical emergency. Richet thought it was due to a weak immune system, but it is now known that it results from a powerful counter-reaction to the toxin by the organism affected. Anaphylaxis (reaction to a toxic substance) was not clearly distinguished from allergy (reaction to a non-toxic agent) until the 1950s.

## The advent of allergies

Around the same time as Richet's discovery, a young Viennese paediatrician called Clemens von Pirquet began to notice the effects on patients of treatment with horse serum, which was injected to give immunity from diphtheria. The reaction to the second vaccination was more sudden and severe than to the first. Pirquet noted extreme feverishness, pains in the joints and a skin rash; he later called the condition 'serum sickness'.

In 1906 Pirquet and a Hungarian colleague, Bela Schick, coined the terms 'allergy' (from Greek *allos*, 'other', and *ergon*, 'reaction') to describe the hypersensitivite reaction they had observed and 'allergen' to denote the substances that caused it. Pirquet foresaw the broad spectrum of ailments that these terms would come to embrace – from hay fever brought on by pollen, to allergic reactions to foodstuffs, to reactions to infectious agents or toxins in the stings of mosquitoes or bees.

## Treatment through desensitising

Pirquet published a comprehensive treatise on allergies. Some, such as hay fever, were attributed to external causes, but no-one had really understood their aetiology: that is, their causes and the factors conditioning their spread. Soon, though, as the incidence of allergies appeared steadily to increase in the world's

**Common allergens**
*An artificially coloured scanning electron microscope image (above) shows allergen-inducing pollen from Bermuda grass (green), ragweed (yellow), maple (red), willow (blue) and plantain (brown). Many allergies are also caused by dust mites (right) – more specifically by their excretions, which are easily breathed in.*

industrialised nations, both the medical profession and the general public in Europe and the USA began to take a greater interest.

Pioneering immunologists had attempted to get to the bottom of the cellular and biochemical mechanisms that underpinned asthma, hay fever and hives, and their work hastened the development of remedies. In 1869, while trying to determine the source of hay fever, the British physician Charles Blackley performed the first skin test by placing pollen on a small break in his skin. Twenty minutes later he observed a rash begin to form. But it was not until 1911 that the British researchers Leonard Noon and John Freeman succeeded in performing the first immunotherapy against pollen-induced allergies. The treatment consisted of injecting ever larger doses of allergens to desensitise the patient, a method that has remained popular right up to the present day.

### TESTING FOR TB

The first diagnostic test for tuberculosis was devised by Clemens von Pirquet in 1907. His skin-test involved introducing subcutaneously a small quantity of liquid tuberculin (a purified extract obtained from a culture of inactive tubercle bacilli) by dropping it onto the forearm and then scratching the skin. If a pronounced local reaction occurred within 48–72 hours of the test, this indicated that the subject had already been infected with the tubercle bacillus. This test was used for decades to diagnose TB and determine the efficacy of vaccines. Other types of testing for TB gauged the reaction of the organism to tuberculin injections, administered either by patches stuck on the skin, a collar of plastic needles that pierced the skin or intradermal injection.

## Relief through antihistamines

In a parallel development, between 1910 and 1920 the British pharmacologist Henry Dale, director of the Wellcome Physiological Research Laboratories, and his colleague, the chemist George Barger, revealed the central role that histamines (chemical neuro-transmitters secreted by certain white blood cells) played in allergies. The first synthetic antihistamine, which blocked particular histamine receptors (known as H1 receptors) was created in 1937 by Daniel Bovet, an Italian pharmacologist working at the Pasteur Institute in Paris. From the 1940s onwards, dozens of derivatives of this molecule began to come onto the market. This new family of revolutionary drugs not only helped to combat allergies but also other ailments such as nausea, psychiatric illnesses and so forth. Even neuroleptic agents such as Largactil® are, in essence, antihistamines.

**Microscopic image**
*Mast cells like this (below), a type of white blood cell, are found throughout the connective tissue of organisms, such as the skin. The nucleus of the mast cell is surrounded by granules that release histamine in the presence of allergens, thus stimulating an allergic reaction.*

### HISTORY OF IMMUNOLOGY

Since ancient times, doctors had noticed that people who recovered from certain illnesses had a natural immunity to them thereafter. Yet immunology as such – the branch of medicine that studies an individual organism's ability to ward off diseases – only really emerged as a scientific discipline in the late 18th century, with Edward Jenner's first vaccinations against smallpox. A century later, the Russian zoologist Ilya Metchnikov showed that certain white blood cells (known as macrophages) have the capacity to absorb and digest microbes and other foreign bodies, thereby giving the organism what is known as 'cell-mediated immunity'. Meanwhile, other researchers such as the German Paul Ehrlich discovered the human body's second great defence mechanism: the humoral immune response. This type of immunity is mediated by antibodies – proteins manufactured by B lymphocytes, another type of white blood cell – which bind to antigens on the surface of invading microbes and neutralise them.

**Pollution in the air**
*Face masks worn to protect against atmospheric pollution in Taiwan. Almost 30 per cent of people worldwide are estimated to be affected by respiratory allergies, compared to 3.8 per cent in 1968.*

# The speedometer 1902

The speedometer was invented at around the same time by the British company Thorpe & Salter and a Strasbourg engineer, Otto Schulze. A flexible cable attached to the transmission (or a front wheel) caused a magnet to rotate. This in turn induced an eddy current in a small aluminium cup (the 'speedcup') mounted above it, which was linked to a pointer. The faster the magnet spun, the stronger the current it generated. The magnetic field pulled the cup and the pointer in the direction of its rotation, indicating the car's speed on a graduated scale.

**Noble marque** *In the Bugatti Type 38 roadster, the speedometer and other instruments were set into a fine walnut dashboard.*

# Disc brakes 1902

Disc brakes were the brainchild of British engineer Frederick W Lanchester. They were based on a simple principle: a disc was attached to the hub of a car's wheel; when the brake was applied, hydraulically operated pads pressed against this disc, slowing the wheel's rotation. But for want of steel that could withstand this kind of treatment, Lanchester's invention lay dormant until the early 1950s. Since then, and particularly as a result of innovations introduced by the Dunlop company, disc brakes have become widespread. They are more efficient and reliable than drum brakes, being less prone, for example, to

deforming and overheating. In the 1980s, composite materials such as carbon fibre and Kevlar brought further improvements in disc-brake performance.

### The man and his marque

Frederick Lanchester (1868–1946) was an automobile and aviation pioneer. In 1896 he built the first practical automobile in Britain, improving it thereafter with innovations such as an improved epicyclic gearbox, fuel injection and turbochargers. The car firm bearing his name was founded in 1900 and went on manufacturing automobiles until 1956.

**Quick stop**
*A disc brake from around 1960, when discs were fast replacing the older drum brake system on cars.*

# Windscreen wipers 1903

**Clear vision**
*Intermittent wipers, with an adjustable delay between wipes, were introduced in 1969. Rain-sensing wipers, which come on automatically and also change speed to cope with heavier downpours, arrived in 1983.*

Windscreen wipers were mainly the work of two American women. In 1903 Mary Anderson devised a basic wiper consisting of two strips of rubber mounted on a pivoting arm, operated manually by a handle inside the vehicle. Before this, when rain or snow fell, drivers simply stuck their heads out of the side window to see where they were going. From 1916, Anderson's wiper became a standard feature on American cars. The next year Charlotte Bridgwood, head of a manufacturing firm in New York, patented her electric Storm Windshield Cleaner. Electric windscreen wipers became commonplace from 1923 onwards.

# A good night's sleep

**B**arbiturates were introduced in 1903 by two German research chemists. The new drug revolutionised the treatment of neurological and psychiatric disorders.

'Ten to fifteen minutes after taking the medication, I fell into a growing state of dejection that led to deep sleep after around 30 minutes. After half a gram of Veronal I slept for 8 hours, and after a whole gram, around 9 hours. On the first morning I awoke fresh and rested; on the second morning, after the higher dose, I found it difficult to get out of bed.' Thus ran the testimony of German medic Hermann van Husen after testing Veronal, the ancestor of modern sleeping pills. The year was 1904 and the young psychiatrist had been commissioned to conduct clinical trials on the drug, which had just been put onto the market by the Bayer pharmaceutical company. An insomnia sufferer himself, van Husen volunteered to take part in the trials of the

**Help to sleep**
*A French advertisement of 1937 for Sédosine, a sedative for the nervous system based on plant extracts.*

**Chemical cosh**
*A mentally disturbed patient and two wardens at an insane asylum in 1891. With the advent of barbiturates, one preferred method of dealing with deeply disturbed patients was to sedate them heavily.*

drug. He was about to test the very first example of what would become an important new family of pharmaceuticals: barbiturates.

## Not just a sedative

The history of barbiturates began in 1864, when the great German chemist Adolf von Baeyer succeeded in synthesising malonyl urea by condensing urea (an animal waste product) with diethyl malonate, which derived from the acid of apples. He called the resulting compound barbituric acid, but it did not appear to have any useful medical properties.

In the early years of the 20th century, Joseph von Mering – a German physician working at the Bayer laboratories – began to take an interest in sleeping aids (hypnotics). At the time, there was little in the way of help for insomniacs. Potassium bromide and chloral hydrate were the most commonly administered treatments; they were effective but had severe

---

### A CALMING INFLUENCE

**B**arbiturates work by producing a depressive effect on the central nervous system. They do this by heightening the excitation threshold of neurones and by lengthening the time they take to return to normal. Barbiturates also enhance the action of GABA (gamma-aminobutyric acid) – the chief inhibitory neurotransmitter in the central nervous system – and block the enzymes that inhibit the synthesis of acetycholine, another neurotransmitter.

---

**Common problem**
*Insomnia affects almost a quarter of the adult population. It is twice as common among women as men and gets worse with age.*

## SLEEP TREATMENT

In 1913 Italian psychiatrist Giuseppe Epifanio used barbiturates to put a 19-year-old girl with severe bipolar disorder to sleep for a two-week spell. After this period of deep-sleep treatment, it was two years before the patient suffered a relapse. The experiment was revived in Zurich in 1920 by the psychiatrist Jakob Klaesi, who established narcotherapy as a way of treating schizophrenia, autism and even delirium tremens. Sleep treatment declined in popularity in the 1930s as it was found to induce cardiac or respiratory problems in around 5 per cent of patients.

**Psychiatric ward**
*An almost empty ward in a mental hospital in Ohio in 1946. While sleep treatment was still being used at this time to treat schizophrenics, the new technique of lobotomy was coming into vogue. This operation involved surgically removing part of the brain. Between 1945 and 1954, it is estimated that some 100,000 people were lobotomised, half of them in the United States.*

side effects. Mering found that these sedatives had a particular chemical structure in common: the presence of two ethyl groups, comprising two carbon and five hydrogen atoms. Sensing that he was onto something, he sought the help of a brilliant colleague, the chemist Hermann Emil Fischer who had studied under Baeyer. By adding two ethyl groups to barbituric acid Fischer created barbital, which the company marketed under the brand name Veronal – a name said to have been suggested by Mering, who thought the Italian city of Verona was the most peaceful place on Earth. The drug soon gained a reputation as an effective sleeping aid and sedative.

Thereafter, with only minimal alterations to its basic chemical structure, chemists have derived some 2,500 compounds from barbituric acid, most famous of which are phenobarbitals such as Gardenal and Luminal (synthesised in 1912). More than 50 of these drugs are in clinical use. The researchers then discovered that these sedatives were equally effective as anti-epileptic agents and as anaesthetics. They proved valuable in the treatment of severe neuroses and psychoses and in other mental illnesses that up to that time had been thought incurable. It was found that a quick-acting intravenous injection of barbiturates could help remove patients' inhibitions and render them susceptible to psychotherapy. This method of treatment grew in popularity after 1940.

### Fatal overdoses

Between 1930 and 1950, barbiturates were prescribed to millions of people worldwide. The dark side of the drugs then revealed itself in severe addictiveness and countless cases of fatal overdoses, either intentional or accidental. The vogue for barbiturates declined after the 1950s following the introduction of new types of drugs: neuroleptics, psychotropic drugs, chlorpromazines and benzodiazepines. Neuroleptic drugs have become particularly useful in modern psychiatry since they are non-addictive. Nevertheless, more than a century after their discovery, barbiturates still play a major role in drug treatment.

# THE AEROPLANE – 1903
# Conquest of the skies

From the first few tentative hops, flying machines developed in leaps and bounds, becoming the most momentous technological advance of the 20th century.

**Gliding German**
*Otto Lilienthal piloting one of his many glider designs in 1891. This influential aeronautical pioneer was interested in bird flight, writing a book on the subject in 1889.*

On the morning of 17 December, 1903, Orville and Wilbur Wright stood on a deserted beach at Kitty Hawk in the US state of North Carolina, gazing in awe at their latest creation, a biplane called the Flyer. The brothers ran a bicycle factory and repair workshop in Dayton, Ohio, but for some time their passion had been the quest to fly. Now, if everything went to plan, one of them was about to take to the air, opening up a whole new chapter in the history of transportation.

The *Flyer* measured 6.4 metres in length and had a wingspan of 12.3 metres. The frame of the aircraft was made from bamboo and walnut covered with doped canvas. Before building it, the Wright brothers had consulted various experts in the new field of aeronautics. One of these was the French-born American engineer Octave Chanute, who lived near

Chicago. Chanute was the author of an influential book called *Progress in Flying Machines*. Published in 1894, this contained all the most up-to-date information on flight and described the work of such aviation pioneers as Sir George Cayley, Otto Lilienthal and the Frenchmen Clément Ader, Alphonse Pénaud and Louis Mouillard.

## Visionaries and dreamers

Sir George Cayley, the English 'father of aviation', was a prolific engineer who was fascinated by the principles of aerodynamics. As early as 1799 he proposed the idea of a fixed wing flying machine and in 1804 he built the first known glider, with large wings towards the front and a small tailplane at the rear. By 1849 he had constructed a biplane glider equipped with 'flappers', which was flown by a ten-year-old boy. In 1853 a glider Cayley developed with engineer Thomas Vick was flown across Brompton Vale in Yorkshire by his grandson George John Cayley. The secret of Cayley's success was his discovery of the four basic aerodynamic forces: lift, drag, thrust and gravity. An understanding of these was crucial to the evolution of powered flight. The first truly methodical investigations into the physics of flight were undertaken by another glider pioneer, the German Otto Lilienthal. From 1891, he built a series of gliders with multiple wings to increase stability. He launched himself in these off platforms and steep hills.

## THE STEAM-POWERED ÉOLE

A rival claimant to the world's first powered flight is the French inventor Clément Ader. On 9 October, 1890, in the park surrounding the Château d'Armainvilliers outside Paris, Ader managed to rise 20 centimetres off the ground for a short hop of 50 metres. His craft was the *Éole*, with bat-shaped wings and a steam engine driving its four-bladed propeller. Ader achieved a few more similarly brief flights in this and a later machine, but his eccentric designs proved uncontrollable. Even so, he was the first to demonstrate that a heavier-than-air craft could lift off from a level surface under its own power. Ader was forced to abandon his experiments in 1897 by his advancing years and lack of funding, but nevertheless he remained a firm advocate of aeroplane development. He famously claimed that 'Whoever gains mastery of the air will rule the world'.

**Model of the Éole**
*Ader's extraordinary bat-winged craft was powered by a two-stroke steam engine of his own devising.*

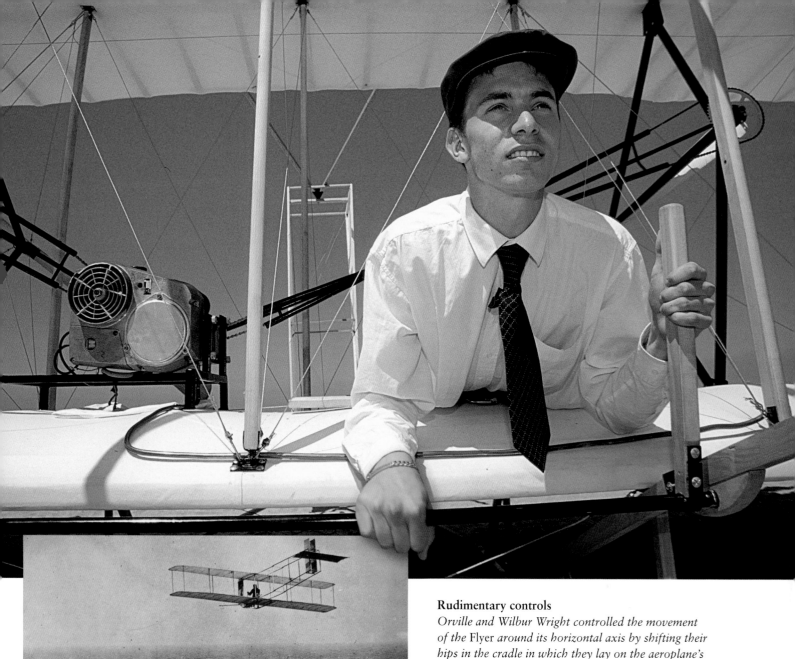

### Rudimentary controls

*Orville and Wilbur Wright controlled the movement of the* Flyer *around its horizontal axis by shifting their hips in the cradle in which they lay on the aeroplane's wing. Here, a modern pilot demonstrates the prone flying position in a replica of their machine.*

## Making the breakthrough

The Wright brothers avidly followed all these early developments in aviation and more, adopting and adapting many of their predecessors' innovations. They chose the largely deserted beach at Kitty Hawk, with its high dunes and strong, reliable winds, as a suitable place to experiment. Throughout 1901 and 1902, the two brothers conducted a series of tests there with unpowered biplanes they constructed. Modifying these machines bit by bit, with different control surfaces, they taught themselves the principles of flying. When piloting their craft, the brothers would lie in a prone position in the centre of the lower wing, facing forwards.

Having failed to find a light enough piston engine to power their machine, the Wright brothers set about making one themselves with

Lilienthal's experiments showed the need to camber the surface of the wings and add stabilisers to them. He also proved that an aircraft's ability to gain lift and glide is directly related to the velocity of the airflow over the wing. Sadly, after completing some 2,000 flights, he was killed in a gliding accident on 9 August, 1896. Lilienthal was on the verge of fitting one his machines with an engine, a propeller and control surfaces.

### Historic site

*The Wright brothers set up a workshop at Kill Devil Hills, above the beach at Kitty Hawk (above), in September 1900. They assembled their first glider there.*

**Wires and struts**
*A model of a pusher biplane,
the MF-11, built by French
aeroplane manufacturer
Maurice Farman in 1913. Biplanes
built by the Farman brothers, Henri
and Maurice, were widely used
for reconnaissance and bombing by
both the French and British in the
early stages of the First World War.*

**Magnificent man in his flying machine**
*Brazilian aviator Alberto Santos-Dumont built this
lightweight* Demoiselle *('damselfly') aeroplane
in France in 1908.*

the help of their mechanic Charles Taylor.
They came up with four horizontal in-line
cylinders weighing 91 kilograms that generated
16 horsepower. A bicycle-chain linked the
engine to two wooden propellers mounted
behind the pilot, who controlled the machine
using a rudder on the tail fin which governed
direction or 'yaw' (that is, swinging to the left
or right) and elevators on struts in front of the
wings which affected pitch (that is, climbing
and descent). Banking the plane into a turn
was achieved through a clever system of wing-
warping: by shifting in his hip-cradle the pilot
deflected the edges of the wings' outer sections
and as the angle of incidence increased on one
side, generating more lift, so the opposite
wingtip would dip, causing the plane to bank.

That first historic flight by *Flyer*, at
10.35am on 17 December, 1903, was made
by Orville. As he began his take-off, Wilbur
ran alongside steadying one wing until the
elevators began to generate lift as the machine
gathered speed. The *Flyer* then left its guide
rails and bounced unsteadily into the air,
flying for about 30 metres before landing
back on the sand. The flight had lasted just
12 seconds. On the second attempt, the plane
made 65 metres, on the third 66 metres and
on the fourth and final flight of the day a full
260 metres. For this last, most successful
flight, Wilbur was at the controls. There were
five onlookers who clapped and cheered as
they witnessed the momentous events.

## Up and away!

Everyone knows the Wrights' epoch-making
achievement now, but at the time it passed
almost unnoticed. The US national press did
not cover the flights at all, while *Scientific
American* magazine merely commented: 'This
represents a step forward in aerial navigation.'

## CROSSING THE CHANNEL

On 25 July, 1909, the French aviator Louis Blériot completed the first aerial crossing of the Channel in a heavier-than-air machine. Flying a monoplane that he had designed and built himself – the Blériot XI, powered by a three-cylinder, air-cooled, 30hp engine – he covered the 24 miles from Calais in just under 38 minutes, crash-landing on the cliff tops near Dover at 5.20 am. Blériot's flight earned him a prize of £10,000 put up by the *Daily Mail* for the first person to cross the Channel by aeroplane.

### French flying flair
*By the time of the First World War, Louis Blériot (right) had produced around 800 of his famous Type XI monoplanes.*

monoplane with a partially metal fuselage. He also patented a means of controlling the ailerons and elevators; movable in all directions, his novel device would later become known as the 'joystick'.

On 13 January, 1908, Henri Farman completed the first officially timed one-kilometre circuit in an aircraft, at Issy-les-Moulineaux. The flamboyant American showman Samuel Franklin Cody was the first man in Britain to conduct a powered flight, on 16 October, 1908. The first Briton to fly was John Moore Brabazon in May 1909 on the Isle of Sheppey in Kent.

The first light and truly reliable aero engine was developed at this time. It was a rotary design, with cylinders set in a radial pattern so they rotated around the crankshaft with the propeller. This configuration helped to prevent overheating and made a heavy water-filled radiator unnecessary. The rotary engine was the

Within a short time, however, aviation would expand from being a pastime for enthusiasts into a mode of transport that revolutionised notions of space and time, radically changing the way in which the world did business.

Others were soon matching the Wrights' achievement. On 12 November, 1906, at Bagatelle just outside Paris, the Brazilian aviator Alberto Santos-Dumont flew for 220 metres at a height of 6 metres in his tail-first biplane, the *14-bis*. This was the first powered flight to be officially recognised by the newly founded International Aeronautical Federation. Press photos of the feat were wired around the world. A few months later, 25-year-old Robert Esnault-Pelterie designed, built and flew a

### WINGS ON WATER

The first seaplane flight was made by a 24-year-old French engineer named Henri Fabre, who on 28 March, 1910, took off from Lake Berre, near Martigues in the Rhône estuary. His aircraft, the *Canard*, was fitted with three floats and powered by a 50-hp Gnome-Rhône engine. Fabre covered a distance of 500 metres at a height of 5 metres before landing safely on the water once more. On 26 January, 1911, the American aviation pioneer Glenn Curtiss fitted floats to a biplane and took off and landed from water, so completing the first successful seaplane flight in the USA. Although they were used in the First World War, the heyday of flying boats was the interwar period, when they were used extensively for long-distance flights. By the mid-1940s they had been rendered obsolete by the development of land-based aircraft that could cross oceans more quickly and cost-effectively.

**Floating flyer** *In the early days of flight the ability to land on water was useful. This German Friedrichshafen FF49 dates from 1917. Today, seaplanes are mainly used where lakes and rivers make it impossible for land-based aircraft to operate.*

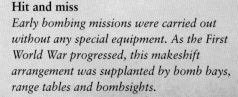

**Hit and miss**
*Early bombing missions were carried out without any special equipment. As the First World War progressed, this makeshift arrangement was supplanted by bomb bays, range tables and bombsights.*

brainchild of a French firm, the Gnome aero engine company. It was a runaway success. Licences for foreign manufacture were sold in Germany, Sweden, the United Kingdom, Italy, Russia and the USA.

## Adopted by the military

The First World War gave an enormous boost to aircraft development as planes were deployed on reconnaissance, bombing raids and pursuit missions. Fighters armed with machine guns became involved in epic dogfights and leading pilots like Albert Ball, Georges Guynemer and Manfred von Richtofen were lionised as gallant air aces.

The French and British manufactured some 125,000 aircraft in the course of the war, while the Germans built around 50,000. By the time of the Armistice in 1918, all the combatant nations agreed on one key point: though air power alone may not have been enough to win the war, any major shortfall in this area would spell certain defeat. The conflict stimulated the development of more powerful aero engines, as well as improvements in power-to-weight ratios, endurance and reliability. By late 1917, fighters like the British *SE5A* and the French *Spad XIII* could reach speeds of 220km/h and could fly at more than 6,000 metres.

## The beginnings of civil aviation

After the war, aircraft development came on in leaps and bounds. In 1919 former military pilots John Alcock and Arthur Brown made the first non-stop flight across the Atlantic in a

**Formidable fighter**
*The* Albatros DV *(top right) was an advanced German fighter introduced in 1916. It saw widespread service on the Western Front, notably with the 'Flying Circus' squadron of air ace Baron Manfred von Richtofen, aka the Red Baron.*

### FIRING FORWARD

No sooner did aeroplanes take on a combat role than engineers began to look for ways of fitting forward-firing machine guns. In front-engined aircraft, this posed a very real risk of shooting off the propeller. French air ace Roland Garros had steel deflector plates fitted on the prop blades, but these caused dangerous ricochets. An interim solution was to mount guns on the top wing of biplanes. In 1915 the Dutch aircraft designer Anthony Fokker, a German resident, devised a synchronised firing mechanism for use on his monoplane fighter, the *Eindecker*. A device known as the interrupter gear paused the gun's firing mechanism each time a propeller blade passed in front of it.

converted Vickers-Vimy bomber powered by Rolls-Royce engines. The following year the French pilot Joseph Sadi-Lecointe became the first to exceed 300km/h; by the end of the decade an Italian, Giuseppe Motta, had almost doubled that speed. The record for a non-stop flight doubled from 5,936km in 1926 to 11,520km by the eve of the Second World War.

The interwar era saw a steady stream of aviation firsts by intrepid pilots such as Charles Lindbergh, Amy Johnson, Charles Kingsford-Smith, Italo Balbo and Amelia Earhart. The period brought significant advances in construction, with the introduction of streamlined all-metal fuselages and enclosed cockpits. Navigation equipment also improved immeasurably. In 1939 Pan American Airways inaugurated the first transatlantic service, with its luxurious *Yankee Clipper* flying boats. Aviation had come a long way since 1903 and would continue to expand, in particular with the introduction of jet airliners in the 1960s that fostered the growth of mass air travel.

**Pride of the Empire** *Imperial Airways was founded in 1924 to provide regular services across Britain's far-flung empire and beyond. Here (above), passengers are boarding a flight at Croydon Airport bound for Paris in 1925.*

**A plane for military and civil use**
*The Lockheed* Constellation *(below) was developed at the instigation of tycoon Howard Hughes and Jack Frye, president of Trans World Airlines. TWA used this fast, long-range aircraft to inaugurate the first regular airline service from New York to Paris in 1946.*

### BIPLANES OR MONOPLANES?

In the early days of flying, the biplane offered a number of advantages over monoplanes. The larger surface area of the wings gave them better lift, as well as making them stronger and able to carry a heavier load. On the debit side, the drag created by the many struts and wires of biplanes made them slower than monoplanes. By the 1930s the introduction of rigid, all-metal cantilevered wings for monoplanes – equipped with extendable flaps that increased the wing's surface area, thus reducing take-off and landing speeds – saw the biplane becoming obsolete.

# The aeroplane

Inventions seldom appear fully fledged out of the blue, and this was especially true of the aeroplane. Many brave and intrepid pioneers risked their lives through trial and error to pave the way for the development of the first successful heavier-than-air machines.

*The parasol wing of the Dewoitine D-27, a French fighter aircraft of the late 1920s.*

## WING SHAPE
### MAXIMISING LIFT

An aircraft wing needs a convex profile to ensure that airflow from the wing's leading edge to its trailing edge creates an area of high pressure on the lower surface, with a corresponding area of low pressure above. The difference between these two constitutes the upward force known as lift. The shape of the wing is a vital factor in how effectively it provides lift. Early aviators laboured long and hard to find the ideal profile. In the 19th century, Englishman Horatio Phillips used a variety of different wing sections to demonstrate that cambered airfoils produce more lift than flat ones. Wing sections remain a focus of much modern research.

## ASPECT RATIO
### REDUCING DRAG

The aspect ratio is the ratio of the overall surface area of an aeroplane's wing to its span. The higher the ratio, the greater the aerodynamic efficiency of the wing, since it reduces drag (air resistance) without impairing lift, resulting in more stable flight. In the early 1900s, some biplanes had wings with a high aspect ratio; however, the advantages gained by the reduction in resistance of the wing were offset by the drag produced by the struts and wires linking the upper and lower wings. It proved even harder to construct a monoplane wing with a high aspect ratio that was light and rigid enough to withstand the aerodynamic loads (compression, twisting, or flexing) that occur in flight. New materials finally allowed construction of wings with ever higher aspect ratios.

*The Blériot XI monoplane had a cambered wing (left).*

*Flaps and airbrakes on a modern commercial airliner (below).*

## THE TAILPLANE
### ENSURING STABILITY

Just as birds have tail feathers, so aircraft are fitted with tailplanes for stability. These usually comprise a fixed vertical surface (the tailfin) to which a movable rudder is attached, plus a horizontal surface for the elevators. The rudder and elevators allow the plane to change direction, controlling pitching and yawing, while the fixed surfaces provide stability in straight and level flight. George Cayley's 1804 glider had a tailplane and in 1899 the Scottish aviator Percy Sinclair Pilcher was killed when the tailplane snapped off his hang glider. Designers still have trouble creating stable 'flying wing' aircraft with no tail.

*The Gloster Meteor (below), the first jet to enter service with the RAF, had a distinctive high tailplane.*

## THE PROPELLER
### OPTIMISING PERFORMANCE

Invented in the 18th century for use in balloons and dirigibles, the propeller was inspired by the sails of windmills. In 1903 Wilbur Wright became the first to realise that it had similar properties to a wing moving through the air. Like a wing, a propeller blade has a particular profile, but unlike a wing the profile changes along the propeller's entire length; together with the vertical plane, this forms an angle of attack known as the pitch. Propeller blades are twisted to achieve a more-or-less uniform angle of attack at any point. Performance was enhanced from the 1930s by the introduction of variable-pitch propellers. These ensure that the pitch is always set at the most efficient angle so that the engine can run at a constant speed regardless of altitude or forward speed. Research is ongoing into high-speed propellers of composite materials to power commercial airliners at up to 800km/h.

*The propeller and engine of the Fokker* Eindecker *monoplane, which entered service in 1915.*

## PISTON ENGINES
### DRIVING THE PROPELLER

One of the greatest challenges for early aeroplane manufacturers was to design and build a suitable engine for use in aircraft. Experiments were conducted with electric motors and even steam engines were tried in the late 19th century, but these proved far too cumbersome. Ultimately the piston engine was adopted with various modifications to reduce weight: for example, the use of aluminium (by the Wright brothers and by Léon Levavasseur in his 'Antoinette' engine of 1903), and the 'V' arrangement of cylinders (Ader, Levavasseur), which dispensed with the need for a heavy flywheel. The first purpose-built light aero engine was the French Gnôme rotary engine of 1908. Its power-to-weight ratio was an outstanding 1hp per kilogram, which over time – through the use of new alloys, better fuels and above all superchargers – was boosted to 1hp per 450g.

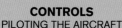

*The 1917 Fokker Triplane, as flown by the 'Red Baron'.*

## RESEARCH AND DEVELOPMENT
### FROM PROP PLANES TO JETS

Frome crude beginnings, the cabin, wings, engines and layout of aeroplanes rapidly evolved. The first all-metal aircraft, the Junkers J-1, appeared in 1915. Aircraft became ever faster through advances in aerodynamics and engine power. Major commercial interest was stimulated by Lindbergh's solo crossing of the Atlantic in 1927. After the Second World War, the development of the jet engine made the piston engine largely obsolete for both military and civil aircraft.

## CONTROLS
### PILOTING THE AIRCRAFT

Pilots use various control surfaces to adjust an aircraft's flight attitude: the ailerons (to bank or roll), the elevators (to climb or dive) and the rudder (to yaw and turn). The Wright brothers developed their first practical flight control surfaces through their use of wing warping. Thereafter, all aeroplanes were fitted with control surfaces, which increased in size as aircraft grew larger. As they did so, it became ever harder to operate the controls manually. In time, aircraft were equipped with mechanical, pneumatic, hydraulic and (from the 1970s onwards) computer-aided (so-called 'fly-by-wire') servo systems to actuate the surfaces.

*An airliner powered by twin turbofans.*

# Chromatography 1903

**Analytical tool**
*A scientist uses compasses on a chromatogram to measure the distance travelled by each constituent element of the substance under investigation (below). This colourised digital image (right) shows the clear separation between the constituents.*

'Like light rays in the spectrum, the different components of a pigment mixture, obeying a law, are resolved on the calcium carbonate column and can then be qualitatively and quantitatively determined.' This is how Russian botanist Mikhail Tsvet attempted to explain chromatography, the method of colour analysis that he had invented. He first presented his findings to scientific peers as early as 1901. In 1903, he published an account of them for wider consumption in *Proceedings of the Warsaw Society of Naturalists*.

## A DELIBERATE PUN?

The term 'chromatography' was coined by Mikhail Tsvet drawing on two Greek roots: *chroma* ('colour') and *graphein* ('writing'). But the word might also be a play on the botanist's own surname, which means 'colour' in Russian. According to this interpretation, 'chromatography' would literally mean 'Tsvet's writing'.

Tsvet was interested in plant pigments. The problem was, having obtained a plant extract, how could it be separated into its constituent elements to facilitate further study? He found the answer quite by chance. Having prepared an extract of spinach with petroleum ether, he filtered the solution by passing it through a column of chalk (calcium carbonate) in a vertical glass tube. As he did so, distinct areas of yellow and green pigment appeared in different parts of the column. Tsvet realised that each pigment had travelled a specific distance before being deposited in the chalk. To obtain pure components of the pigment, all he had to do was to take samples from each colour zone. This was the basic principle behind the science of chromatography.

## Neglect and revival

Tsvet's work excited some interest, but it was soon forgotten in all the upheaval of the First World War and the Russian Revolution. Tsvet died in 1919, aged just 47. Then, in 1931, two biochemists at the University of Heidelberg, Edgar Lederer and Richard Johann Kuhn, were conducting their own research into plant pigments and resurrected his technique. Several new methods of chromatography were devised thereafter along essentially the same lines: that is, by filtering a liquid or a gaseous compound through a medium – paper, for example, or a porous material, a gas, an immobilised liquid – that retains each of the separate components at a particular level.

Chromatography has since become an indispensable tool in organic and biochemical research. It is used, for example, to detect drugs in athletes' blood samples; to isolate a particular ingredient for drug manufacture; and more generally to separate, analyse and identify different elements within compounds.

# Monopoly 1904

In 1904 Elizabeth Magie Phillips, an American board-game designer, was granted a patent for 'The Landlord's Game', which she created to demonstrate the evils of property speculation. As a follower of the radical economist Henry George, Phillips wanted her game to show that renting only served to enrich landlords and to make the poor even poorer. The game became extremely popular by word of mouth and its rules evolved accordingly.

In 1933 an unemployed former salesman called Charles Darrow, who was a devotee of 'The Landlord's Game', decided to customise it with street and property names from Atlantic City, New Jersey. After trying out his 'new' game with friends and acquaintances, he began marketing it under the name 'Monopoly'. It proved an instant hit in the feverish atmosphere following the Wall Street Crash of 1929.

Parker Brothers acquired the patent to Monopoly from Darrow in 1935. The following year they also recognised Phillips' part in its invention by paying her an honorarium of $500. By the 21st century, more than 200 million Monopoly sets had been sold across 80 countries, and the game is produced in 26 different language versions.

**Big board game**
*In 2005 Parker Brothers marked their 70th Monopoly anniversary by producing the largest ever board, measuring 440m².*

# Offset printing 1904

In 1904 Ira Washington Rubel, a New Jersey printer, forgot to load a sheet of paper into a lithographic press he was operating. As a result, the rubber roller that was designed to press the paper down firmly on the inked printing plate roller picked up the image instead. When the machine was next run, with paper in it, the paper emerged printed on both sides: the right way up by the roller on which the printing plate was mounted and upside-down by the rubber-covered impression cylinder. The surprising thing to Rubel was that the image from the rubber impression cylinder was clearer. He had accidentally discovered the principle of offset litho printing.

The lithographic printing process is based on the mutual repulsion of waxy ink and water. By adding a third roller – the blanket cylinder – between the plate roller and the impression roller, the quality of the printed result was greatly improved. The paper no longer came directly into contact with the printing plate cylinder, since the image was first transferred – 'offset' – onto the blanket cylinder, and it was from this that the impression was taken. Offset lithography became the most widespread form of printing from the 1960s onwards.

**Modernity in print**
*In 1926 typographer Joost Schmidt designed this handbook on offset printing for the Bauhaus, Germany's famous art and design school.*

---

**THE LITHO PLATE**

In offset lithography, the text or images to be printed are transferred photographically onto a printing plate made of thin aluminium or other flexible material, such as polyester or mylar.

---

# The body's chemical messengers

From the mid-19th century onwards, the discovery by researchers of different glandular secretions in the human body brought to light the existence of what would later become known as hormones, and hinted at the different functions they performed. Endocrinology and physiology were revolutionised by these findings.

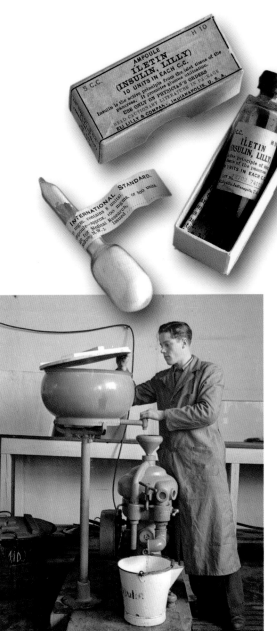

Hormonal activity was first identified by researchers in the 1850s. The Japanese chemist Jokichi Takamine isolated the hormone adrenaline from animal glands in 1901. The term 'hormone' – from the Greek verb hormâne, meaning 'to excite or wake' – came later, being coined in 1905 by the physiologist Ernest Starling in lectures given at the Royal College of Medicine in London.

In 1855 the British physician Thomas Addison published a paper on chronic diseases of the adrenal gland, 'On the Constitutional and Local Effects of Disease of the Suprarenal Capsules'. This seminal work inaugurated the discipline of endocrinology, the study of the function and disorders of glands. Around the same time, the German Arnold Berthold's experiments on roosters revealed the key role that gonads (male sex glands) played in the development of secondary sexual characteristics. Berthold concluded that there was a substance produced by the testes and released into the blood stream that acted upon the whole organism.

**Relief for diabetics**
*Canadian researchers Frederick Banting and Charles Best were the first to isolate insulin, in 1921. Two years later, it went on sale in the USA under the name 'Iletin' (above). This worker at a pharmaceutical plant at Bitterfeld, East Germany (right), is extracting insulin from the pancreas of animals in 1946.*

### Glands as the elixirs of life

In 1899 Harvard physiologist and neurologist Charles-Édouard Brown-Séquard undertook the first clinical trials of hormone therapy by injecting himself with an extract from dog and guinea pig testicles. Aged over 70 at the time, Brown-Séquard recorded feeling mentally and physically reinvigorated by his 'rejuvenating elixir'. His hormone replacement method gained further acclaim when a patient suffering from hypothyroidism was successfully treated with extracts from the thyroid gland. By the end of the 19th century, testosterone was being prescribed for a wide range of illnesses.

In London, Ernest Starling and his brother-in-law William Bayliss were studying the nerve supply to the small intestine. Experiments conducted by Russian physiologist

## A MEDICAL BOON

As well as performing a vital role in the normal functioning of the body, hormones are prescribed to counter underproduction by glands or to alter hormonal cycles (for example, oestrogen in the contraceptive pill). Large-scale commercial production of hormones began in the 1950s. Originally, they were derived mostly from animal glands or from glands taken from human cadavers. Thanks to recent advances in gene research, laboratories can now manufacture artificial hormones which pose no risk of contamination and are also more suitable for treatment of humans than hormones from other animals. This should help to avoid tragic incidents such as occurred in the 1980s, when human growth hormone contaminated with the agent responsible for causing Creutzfeldt-Jakob ('Mad Cow') Disease led to deaths of hundreds of people.

Ivan Pavlov some years before had established that the digestive process was controlled by a nervous reflex induced by hydrochloric acid. Mainstream medical thinking at the time held that all the major functions of the body were governed by the nervous system. Starling and Bayliss exploded this theory. After severing the nerves linking a dog's duodenum (the first part of the small intestine) to its pancreas, they instilled a weak solution of hydrochloric acid into the duodenum; despite the lack of nerve endings, secretion from the pancreas still occurred, indicating that something other than a nervous reflex must be at work.

## Secretin – a chemical messenger

In an experiment in 1902 Starling and Bayliss scraped mucous from a dog's duodenum, ground it up with sand in a mortar and added hydrochloric acid. After filtering, this mixture was injected into the veins of another animal, where it instantly stimulated the production of pancreatic juices. This experiment proved conclusively that a duodenal hormone carried by the blood was what regulated the secretion of pancreatic juice. The scientists called it secretin.

In fact, secretin turned out to be just one of a host of hormones that were found to act as chemical messengers between organs. From the 1920s the functions of the endocrine glands (the pineal, pituitary and adrenal glands, thymus, pancreas,

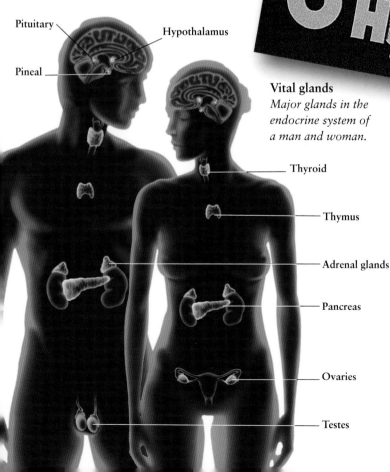

**Vital glands**
*Major glands in the endocrine system of a man and woman.*

Pituitary

Hypothalamus

Pineal

Thyroid

Thymus

Adrenal glands

Pancreas

Ovaries

Testes

**Enhanced virility**
*The testosterone-based drug Sterandryl® went on sale in the 1950s as an anabolic steroid, but adverse effects soon forced its withdrawal.*

ovaries and testes) were established and the hormones that they secreted were identified. Gradually scientists uncovered the relationship between these secretions and the role played by the pituitary gland in stimulating hormone production. (The pituitary is in the brain where it is closely linked with the hypothalamus, which triggers many of its actions.) Hormones produced by the thyroid control important bodily functions such as the consumption of energy, cardiac rhythm and body temperature. The parathyroid hormone (PTH) controls calcium and phosphorus levels in the blood, ensuring healthy development of bones.

# New music for a modern age

**A**s the 20th century dawned, a new kind of popular music arose in the United States. It was an infectious blend of African and American idioms played predominantly by black musicians, who turned the musical world on its head with their skilful improvisation and syncopated rhythms. Before long it was known all around the globe as 'jazz'.

As a fusion of the Western musical tradition with the music of the descendants of black slaves, jazz came to epitomise the cultural melting-pot that was American society. For Africans sold into slavery and transported to America, music became an outlet for their suffering and sorrow and an expression of their cultural identity. Using instruments associated with the European classical repertoire – notably the piano, trumpet, clarinet and saxophone – black musicians created a wholly new idiom characterised by 'swing', a fluid style of playing with a strong underpinning rhythm that owed much to the dance and ceremonial music of their African forefathers.

Similar cross-fertilisation gave rise to the Negro spiritual and Gospel music, in this case a blend of the Protestant choral canon and African song, and also to the Blues, an improvisational style of music expressing melancholy and sadness, which arose from the work chants of plantation slaves and the tradition of oral storytelling. These musical styles enjoyed instant popularity. Inspired by this rich heritage, musicians like Jelly Roll Morton on piano and Sidney Bechet on clarinet and saxophone created the form of music that came to be known as 'jazz'.

### Improvisation is king

In contrast to Classical music, where players strive to play exactly what the composer has written in the score, jazz is improvised.

**BLUE NOTES**

**I**n jazz, blue notes are notes played or sung a semitone lower than the pitch of the major scale. They are often called the flattened third, flattened fifth and flattened seventh. The name comes from the fact that songs using these evocative notes were commonly sung at dusk, after work, by slave gangs. By extension, the term came to be applied to a whole sub-genre of jazz.

Working from an opening theme – usually a well-known melody plus a few chords – jazzman wove an infinitely varied tapestry of sound. Taking cues from one another, players had endless freedom to create new melodic phrases and rhythms. Above all, the key requirement for a jazz musician was to be able to improvise. Many players became renowned not only for their virtuosity on an instrument but also for their improvisational ability to come up with original phrasing and breathtaking variations on a central theme.

Likewise, unlike the sweet harmonies of Classical music, in Gospel and Blues both the instrumentation and the voice were used for expressive ends, to convey raw emotion.

**All present and correct**
*Louis Armstrong's famous band the 'Hot Five' in its 1925–6 line-up comprising (left to right): 'Satchmo' on trumpet (seated at the piano), Johnny Saint-Cyr on banjo, Johnny Dodds on clarinet and alto sax, Kid Ory on trombone and Lilian Hardin (Armstrong's wife from 1924 to 1932) on piano.*

A prime example is the warm, gravelly voice of Louis Armstrong ('Satchmo') or the rich tones of his trumpet, which he would often vary by using a mute. In the same way, jazzmen were quick to adopt the microphone to amplify their sound. Audiences unused to such full-on, powerful delivery were shocked at first, but jazz quickly grew in popularity.

## An infinite variety of forms

The cradle of jazz was the city of New Orleans in Louisiana. In the 1910s, the predominant style of music played there was characterised by the swaying rhythm of ragtime (a dance form created by African-Americans) and the harmonies of Blues. Key figures in the rise of New Orleans jazz were black and creole musicians like the cornet players Buddy Bolden and Sidney Bechet. They fronted small ensembles of three to six instruments, with the saxophone or trumpet as the lead supported by a rhythm section comprising a tuba or double-bass, plus a banjo or piano. These jazz combos were also responsible for the development of the modern drum kit.

Like all popular music, jazz evolved to fit in with prevailing tastes and fashions. From

### WHY 'JAZZ'?

Nobody knows exactly where the term 'jazz' came from. Its roots may go back to Africa – to the Bantu term *jaja* ('to play music'), for example, or the adjective *jasi* ('excited'). Some commentators trace it to the French verb *jaser*, meaning 'to gossip', while others think it may be a shortened form of 'jasmine'. What is beyond dispute is that it first appeared in the New York press in 1917, in the name of a group called the 'Original Dixieland Jazz Band'.

**In the swing**
*A French poster issued by His Master's Voice in 1920. Record companies wasted no time in capitalising on the craze for jazz.*

**Virtuoso player**
*Sidney Bechet playing clarinet for a recording session in the Blue Note studios on 8 June, 1939. His version of Gershwin's 'Summertime' was the first hit for the New York label, founded by two German-Jewish immigrants, Francis Wolff and Alfred Lion.*

**Jazz Age**
*The central panel of a triptych called* Großstadt *('The Big City) made by the artist Otto Dix in 1927. The painting depicts patrons of a Berlin cabaret club dancing to a jazz band in Weimar Republic Germany.*

**Lady sings the Blues**
*Billie Holiday (below left), whom saxophonist Lester Young nicknamed 'Lady Day', was discovered by a producer of Columbia Records. She sang with all the jazz greats of the big-band age: Benny Goodman, Duke Ellington, Artie Shaw and Count Basie. The clear yet hoarse tone of her voice and her drawling delivery are instantly recognisable. She modelled her singing style on the playing of jazz instumentalists.*

New Orleans it spread north to the clubs of Chicago and New York, fertile seed-beds for further development. As it became more established, jazz entered its classic phase, heard in the music of Louis Armstrong, Jelly Roll Morton and King Oliver. Advances in recording techniques meant that their performances were captured for posterity and heard by millions via records and radio.

In the 1920s and 1930s, a new generation of jazzmen emerged, led by the likes of Duke Ellington, Benny Goodman and Lester Young, whose playing was more formal and structured than that of their predecessors. This phase saw the rise of the big band sound. After the Second World War jazz went global, spread by US servicemen and profiting from the popularity of all things American after the liberation of Europe and the Far East. It even influenced a new breed of classical composers and writers of film scores.

During the 1950s, jazz grew ever more diverse with the emergence of radically different styles such as bebop (fast, virtuosic playing

## JAM SESSIONS

It may sound as though there are no rules, but before launching into an improvisation, jazz musicians generally choose a central musical theme (the 'standard'), comprising either a specially written melody or one borrowed from the folk, pop or classical repertoires. The piece then develops according to a principle of alternating passages by the ensemble and the soloist, who provides the 'chorus' (an improvised solo). To help guide the rhythm section, the succession of chords on which the piece is based is set down in the form of a chord chart (with each chord corresponding to a bar of music). This harmonic and rhythmic 'ground' can be repeated ad infinitum, though jazz musicians rarely follow this formal framework to the letter.

## THE BIG APPLE GETS JAZZ

**J**azz arrived in New York relatively late, only really gaining a foothold there in the late 1920s. But once Duke Ellington's orchestra had taken up residence at the Cotton Club (which still had a colour bar in place against African-American customers) and Chick Webb at the Savoy Ballroom, many others followed. Jazz combos such as those of Fletcher Henderson, Count Basie, Jimmie Lunceford and Lionel Hampton appeared regularly at the city's Apollo Theatre, as New York developed a thriving jazz scene. From the 1940s onwards, clubs on 52nd Street (nicknamed 'Swing Street' and 'the street that never sleeps') played host to the new generation of jazzmen who invented bebop.

**The Count Basie Orchestra**
*One feature of Basie's big band (above) was the sound of 'duelling' between virtuoso solo instrumentalists, famously between tenor sax players Herschel 'Tex' Evans and Lester Young. Evans died in 1939, aged just 29, before this photograph of the orchestra was taken in New York in 1944.*

with ever-changing, complex rhythms); leading exponents of bebop were Charlie Parker on sax, Theolonius Monk on piano and the singer Sarah Vaughan. The style known as 'hard-bop' was even more radical, reintroducing church and gospel elements back into jazz.

One remarkable player who mastered both the hard-bop form and the far more mellow, laid-back style of 'cool jazz' was the trumpeter Miles Davis. In the 1960s innovators such as the multi-talented Ornette Colman discarded fixed chord changes and tempos altogether to create so-called 'free jazz'. The following decade saw attempts by some jazz musicians, notably Miles Davis, to bring elements of the dominant popular idiom of rock into their music, so giving rise to 'jazz fusion'.

**Jazz aristocrat**
*Pianist, composer and band leader Duke Ellington, seen here (left) in 1934, formed his first orchestra in 1923. Four years later they were playing at Harlem's famous Cotton Club.*

# Mastering the space-time continuum

**Young Einstein**
*A portrait of Einstein taken in 1902, when he was working as an assistant examiner of patent applications in Bern, Switzerland.*

On 30 June, 1905, Albert Einstein published a paper in the journal *Annalen der Physik* outlining what came to be called his 'special theory of relativity'. The young physicist was still employed at the Swiss patent office, but he was about to take the scientific world by storm.

**Future vision**
*The theory of the expansion of time has been fertile ground for science-fiction writers and film makers. Many works have probed the space-time question, notably the 1968 film 2001: A Space Odyssey (left), directed by Stanley Kubrick.*

The truly startling conclusion in Einstein's 1905 paper 'On the Electrodynamics of Moving Bodies' was that the passage of time, hitherto thought immutable and universal, is in fact relative to the movement of the person measuring it. If one imagines, for example, a video conference between a person travelling in a spaceship and another on Earth, then the latter will see the former speaking and moving in slow motion. Strange though this sounds, the effect has now been observed and verified.

## Challenging Newtonian physics

Einstein's paper marked the beginning of a fundamental shift in our understanding of the universe. Newton's law of universal gravitation had reigned supreme in physics since its publication in 1687. Yet it had been apparent for some time that there was a major flaw in Newton's grand theory: in some circumstances light did not appear to obey his rational principles, but instead behaved in an odd and

---

### $E = mc^2$

In October 1905 Einstein submitted an addendum to his original paper, reflecting a new finding that had just emerged from his theory: that matter and energy (in the form of electromagnetic radiation) are two aspects of the same physical entity, as reflected in his equation $E = mc^2$ (E being energy, m mass and c the speed of light). It is evident from, say, a bomb exploding that matter can transform itself into energy, but that energy – a ray of light – could manifest itself as a physical body was a remarkable new conclusion. Experiments in particle physics have since shown Einstein to be right.

**Time tunnel**
*This structure (left) was built in Seoul (South Korea) in 2005, as part of an exhibition marking the centenary of Einstein's special theory of relativity.*

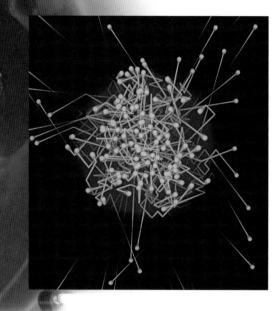

**Light display**
*The Sun emits light in all frequencies: visible, infrared and ultraviolet. This image (above) shows the paths of photons as they fly away from the Sun.*

erratic way. If, for example, a passenger in a train travelling at 100 metres per second (or 360km/h) and a person standing on a railway platform could both observe the same particle of light (photon), they would see it moving at the same speed, despite one person being in motion and the other being stationary. By Newton's law, which seemed the more rational position, the opposite should be true.

Let us now imagine that the experiment is repeated with a balloon rather than a light particle. The man on the train launches the balloon at a speed of 10 metres per second (m/sec) into the direction of travel. The man on the platform sees the balloon moving at a speed of 10m/sec, plus the 100m/sec that the train is travelling – in other words a combined speed of 110m/sec (or 396km/h). This appears to be simple common sense. And the fact remains: if the balloon were a photon, the two observers would measure it at exactly the same speed – 110m/sec (see panel, right).

This was a real conundrum. Between 1881 and 1887, experiments conducted by the American physicist Albert A Michelson had demonstrated beyond doubt that light always travels at 300,000m/sec irrespective of the motion of the observer. The implications of his finding were immense: if light was exempt from Newton's laws, then this undermined the very foundations of physics. For this reason, most scholars at the time simply ignored Michelson's findings. Einstein, meanwhile, was struggling to find a way to resolve the paradox. The only solution he found that worked was to abandon the notion of time as an immutable constant. He was forced to conclude that time did not, in fact, pass at a uniform rate everywhere.

## A QUESTION OF TIME

Consider the experiment involving a train described in the main text (left). Imagine that, to measure the balloon's speed, each observer is given a stop-watch and a ruler. An object's speed can be determined by measuring the time it takes to cover a certain distance. If it travels 10 metres in 1 second, its speed is 10m/sec. If we postulate that the stop-watch used by the man on the train is running slower than that of the man on the platform, then the paradox of the constant speed of light can be resolved. The man on the platform sees the balloon travel 110 metres in 1 second, whereas the man on the train – whose watch has only moved on one-eleventh of a second (0.09sec) – sees the balloon cover 10 metres. And 10 metres in 0.09 seconds equates to 110 m/sec. In other words, the two observers arrive at the same result.

## PROOF POSITIVE

The first concrete proof of the theory of relativity came on 29 May, 1919, during a total eclipse of the Sun in the southern hemisphere. Observing this event from the island of Principe was British astronomer Arthur Eddington. He demonstrated, with photographic evidence, that the stars closest to the Sun appeared to have shifted in relation to their actual position, as a result of the deflection caused by the mass of the Sun. Proof of Einstein's assertion that time slows down with increased acceleration had to wait until the 1950s and the advent of atomic clocks: a clock on board an aircraft circling the Earth and another in Washington registered a difference of 59 billionths of a second.

**Genius at work**
*Einstein at the blackboard during one of a series of lectures he delivered in Paris during the spring of 1922.*

Virtual position of the star as seen by an observer on Earth

Actual position of the star

Sun hidden by a lunar eclipse

Path of light

Observer on Earth

**Not an optical illusion**
*As particles without mass, the photons that make up the rays of light emitted by stars are not subject to attraction by the Sun. This meant that, in Eddington's eclipse experiment (above), curvature in space itself must have caused the positions of the stars to shift.*

## An absolute limit

To return to the photon/balloon experiment once more, if we posit that time passes more slowly for the man on the train than it does for the one on the platform, the problem is resolved. Yet this is not the end of the matter. If we accept Einstein's account, one of the upshots is that the speed of light (c) has an absolute limit that no matter (or energy) can exceed. Indeed, according to the equations that Einstein formulated to encapsulate his theory, including $E = mc^2$, the closer one comes to the speed of light, the more time slows down. And once arrived at the point c (terminal velocity), time would cease to flow and the particles of light would exist in an eternal present.

Little wonder that his contemporary scientists viewed Einstein's proposal with scepticism. Yet his theory of special relativity ('special' in the sense that it only applied to uniform motion in a straight line) soon proved unassailable. And as experimentation confirmed it, so it became established as a new fundamental law of physics.

## The general theory of relativity

From late 1905 onwards, Einstein began to investigate a far more difficult problem: accelerated motion. In 1907 he had what he

## UNIFORM AND ACCELERATED MOTION

An object moving in a straight line at a constant speed is said to be in uniform motion. If its velocity alters, either in magnitude or in direction (or both), its motion is described as 'accelerated'.

*The importance of Einstein in modern-day physics cannot be overstated. His work revolutionised ideas of space, time, matter and energy and our entire understanding of the universe. Quantum mechanics is continuing where he left off.*

**Relative motion**
*Due to the effects of gravity, for the same amount of energy expended, this long-jumper would travel several metres further on the Moon, while on Jupiter, he would barely manage a single metre.*

described as 'the happiest thought of my life,' when he realised that acceleration and gravity are equivalent – they are indistinguishable from one another without a frame of reference. This insight paved the way for his general theory of relativity, published in 1915, which stated the revolutionary idea that an object resting on the ground and one that is accelerating are subject to one and the same physical force.

## Galileo in a lift

Einstein had reached his idea through a thought experiment involving two elevators. With impeccable logic he imagined two people, each enclosed within an elevator with no view of the outside world. One elevator was on terra firma

### EINSTEIN'S NEW COSMOLOGY

In a paper published in 1917, Einstein applied general relativity – a theory hitherto confined to explaining the particular matter of masses in motion – to the universe as a whole. His proposal that the universe was spherical paved the way for the study of the structure and evolution of the universe, which in turn revealed that its size was neither infinite nor constant. In the 1920s, hypotheses and proofs that the cosmos was constantly expanding gradually led to the 'Big Bang' model. According to this, some 13.7 billion years ago the universe exploded from an unimaginably dense and hot primordial condition. The Big Bang theory was first advanced in 1927 by the Belgian priest and mathematician Georges Lemaitre, who called it the 'hypothesis of the primeval atom'. The idea was championed and developed by the Russian-born American physicist George Gamow from 1948 onwards.

**Black holes** *The black hole phenomenon results from a giant deformation in the space-time continuum, usually caused by the explosion of a star. The resulting gravitational field is so strong it deflects the path of all surrounding light and matter to the point of absorbing it.*

Tower of Pisa) were enclosed within the accelerating compartment, there would be no reason for his cannonballs, once released, not to hit the floor at the same time, since in the absence of gravity it would not be the balls that were falling but the floor that was accelerating up to meet them.

## Time and motion

Einstein drew some extraordinary conclusions from his general theory of relativity. One was that time slows down as an object accelerates. In the original train/balloon experiment, acceleration was not at issue; the train was moving at a constant speed of 360km/h, and the observer on the platform ascertained that time on board the train passed more slowly than it did for him. But now Einstein was able to complete the picture. When the train is standing in the station, the two men experience the same time. But as the train pulls away and accelerates to reach its cruising speed, time begins to slow down for the man on board.

## GRAVITATIONAL LENSES

The most spectacular proof of general relativity is the effect of gravitational lenses, observed for the first time in 1986. These arise from the warping of space-time around a massive object – generally a cluster of galaxies – and bend light rays as though through a gigantic glass lens. Astronomers have used this phenomenon to peer into deep space. In 2004, for example, the Hubble Space Telescope utilised the magnifying effect produced by the cluster Abell 2128 to increase its power of resolution 25 times and observe a galaxy 13 billion light years away. The same year, the European Space Agency's VLT telescope, using Abell 1835, saw a galaxy 13.23 billions light years away.

while the other was moving through space with an acceleration equal to one Earth gravity. He claimed that the people would have no way of telling whether their weight, which was what was keeping their feet planted on the floor of the compartment, was due to the downward force of gravity or to the upward force of acceleration.

This discovery finally resolved a 300-year-old problem that had been troubling scientists ever since Galileo dropped his cannonballs from the Leaning Tower of Pisa. Galileo had surmised that two balls of different weights (masses) will fall at the same rate and hit the ground at the same time. But despite proving this by experiment, it still seemed illogical: why didn't the heavier cannonball fall faster? To provide an answer, Einstein applied his hypothesis that gravity equals acceleration.

To start with, Galileo's experiment (which had been conducted in Earth's gravitational field) would have the same outcome if it was done in an enclosed compartment accelerating upwards through space. In this latter case, the apparent paradox of two bodies of dissimilar mass falling at the same rate would simply disappear: if Galileo (and with him the

**Einstein Cross**
*The best-known example of a gravitational lens is the Einstein Cross (right), resulting from four symmetrically placed images of the same quasar (the energetic core of a remote active galaxy). The focusing effect is produced when a concentrated mass (galaxy cluster) lies between the quasar and Earth.*

## The twin paradox

Another startling conclusion was illustrated by the so-called 'twin paradox'. This thought experiment involves identical twin brothers. One of the brothers is sent into space and spends a year on board a spaceship that accelerates almost to the speed of light. Because the acceleration slows down time for the astronaut, on his return he finds that his twin brother and everything else on Earth have aged whereas he has not. Langevin resolved this apparent paradox by demonstrating that time dilation is asymmetrical. But there is something stranger still going on. As acceleration slows down time, this can be directly transposed to gravity via the equation 'gravity = acceleration', so that the stronger the force of gravity (when close to a black hole, for example) the slower time passes.

## Warped space

Einstein's final conclusion was that space is warped by the mass of celestial bodies within it. Flying in the face of the usual concept that space is three-dimensional, which would deny it the ability to bend, Einstein took the radical step of thinking of space as two-dimensional – like a stretched piece of cloth in which billiard balls (planets and other objects) form hollows. It is this characteristic that explains how photons (particles with no mass) appear to be deflected by gravity, which by definition only acts on bodies with mass. The phenomenon is in fact due to the warping of space: the photon behaves like an ant walking around the edge of a bowl, all the while imagining it is moving in a straight line. Einstein's theory completely altered our view of time, space and the universe.

**Brothers in science**
*The French physicist Paul Langevin, a disciple of Einstein – seen here (above, on the left) with Einstein in 1922 – formulated the 'twin paradox' to elucidate the general theory of relativity. Langevin was also known for his work on sonar during the First World War.*

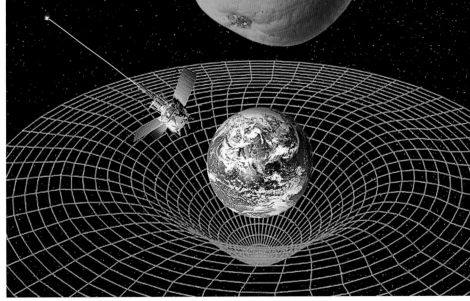

**Paradigm shift** *Newtonian physics regarded gravitation as a force that was responsible for objects falling (like the orange above) and also for the attraction between planets. Einstein, in contrast, saw it as a deformation in space-time produced by mass and the rotation of the Earth. This CGI image is based on data collected by Gravity Probe B, which was used by NASA until 2005 to verify Einstein's assertions on cosmology arising from the general theory of relativity.*

# The archetypal genius

I n the popular imagination, Einstein was a maverick thinker who turned the laws of physics upside-down while sticking out his tongue at the world. But the man behind this mischievous mask was a mass of contradictions. Before publishing his groundbreaking papers on physics, at the age of just 26, he had been an underachiever both at school and college, a mediocre student who never did any practical experiments. His initial theory was based largely on intuition.

'The contrast between the popular assessment of my powers and achievements and the reality is simply grotesque', wrote Albert Einstein in 1931. Perhaps the great thinker had some foreboding of the macabre scenes that the genius cult surrounding him would provoke following his death on 18 April, 1955. No sooner had the 76-year-old physicist passed away at Princeton Hospital, New Jersey, than the pathologist, Dr Thomas Stoltz Harvey, removed his brain during autopsy, without permission, in the hope of discovering what it was that made Einstein so extraordinary. As it turned out, the brain weighed 200 grams less than average. Its only unusual feature was that the left hemisphere was more highly developed than in most people. This was to the detriment of the part of the brain which controls language, which is consistent with the fact that the young Einstein was a slow developer, only acquiring speech relatively late in childhood.

Einstein was born on 14 March, 1879, to a non-practising Jewish family in the southern German city of Ulm. From an early age, he

**College sweetheart**
*Einstein's first wife was Mileva Maric, a young Serbian mathematician. She is pictured here in 1914 with their sons Eduard (left) and Hans Albert.*

**Into exile**
*Einstein playing the violin on his return voyage to Germany on the liner* Belgenland *in early 1933 (top). When he learned that Hitler was the new German Chancellor, Einstein got off the ship at Antwerp and went back to the USA, where he stayed for the rest of his life.*

displayed great powers of concentration and an aptitude for original thinking. As his sister Maja recalled 'even when there was a lot of noise, he could lie down on the sofa, pick up a pen and paper... and engross himself in a problem so much that the background noise stimulated rather than disturbed him'. At the age of 5, he was already showing more interest in the compass owned by his engineer father Hermann than in the violin lessons he received from his mother Pauline. Albert was a solitary and quick-tempered child, who later learned to adopt a dreamy air of detachment to anything that threatened to disturb his work. He formulated his Theory of Special Relativity by thinking up imaginary scenarios and devising thought experiments, in the mould of Galileo and Isaac Newton, two of the scientists he most admired.

### Self-taught physicist

After enrolling at the Zurich Polytechnic Institute in 1896, the 17-year-old Einstein began to come out of his shell. During his time there he met Mileva Maric, a brilliant mathematician, whom he married in 1903. Three years earlier he had passed his exams in mathematics and physics, but it was largely through his own efforts that he acquired a thorough knowledge of contemporary developments in physics. He struggled to make ends meet during this phase of his life, but it was also a time of great intellectual endeavour and achievement for him.

In 1901 Enstein took up the post of technical assistant (third class) at the Swiss Federal Patent Office in Bern, a lowly position that happily left him plenty of time to think and pursue his own scientific investigations. Just four years later, as a complete unknown, he published his article on special relativity, which stirred up the learned world of physics

**US citizen**
*On 1 October, 1940, Albert Einstein, his secretary Helen Dukas (left) and his stepdaughter Margot (right) swore the oath of allegiance to the United States of America and became naturalised US citizens.*

and laid the foundations for quantum theory. There followed, in 1915 and 1917 respectively, works on the general theory of relativity and on its application to cosmology.

It took time for Einstein's theories to be widely accepted: even the Nobel prize that he was awarded in 1921 made no mention whatsoever of relativity. In an attempt to win over reluctant fellow scientists, Einstein embarked on an extensive lecture tour around Europe in the 1920s. It was at this time that his work first came under attack by anti-Semitic elements, who decried his theories as 'degenerate' and tried to smear him with accusations of plagiarism. But by now his reputation was too firmly established.

## The A-bomb and quantum physics

With Hitler's rise to power in Germany in 1933, Einstein went into exile in the USA, taking up a research post at the University of Princeton. In 1939, despite his fundamentally pacifist beliefs, he urged President Franklin Roosevelt to develop the atomic bomb for use against the Nazis. But by 1945, he regretted his involvement in its development. On the eve of the bomb's deployment against Japan, he begged President Truman to think again.

Einstein was also at the forefront of quantum physics in the 1930s, even though he rejected the uncertainty principle at its heart (formulated by German physicist Werner Heisenberg) with the famous statement 'God does not play dice'. After the war, Einstein stepped back from the limelight but continued his researches. At his death, he was working on a Grand Unified Theory that would embrace all the fundamental laws of physics. His successors are still engaged on that quest.

**Captured for posterity**
*This famous photo of Einstein was taken on his 72nd birthday on 14 March, 1951.*

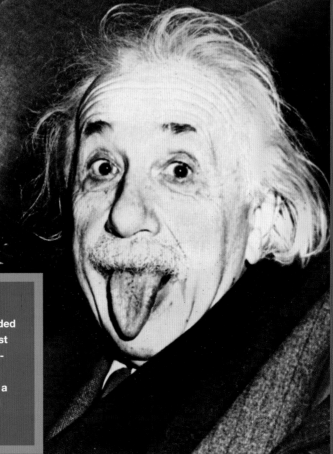

### A CULT FIGURE

Everyone knows the famous picture of Einstein the tousle-headed iconoclast, poking his tongue out at the camera. This was just one of many faces of a man who, by turns, epitomised the absent-minded professor, the melancholic loner, the laughing genius and the selfless scientific pioneer. Einstein became every bit as much a universal archetype as Mahatma Gandhi in the political arena and Pablo Picasso in the world of art.

# Sonar 1905

On 9 January, 1917, Kaiser Wilhelm II of Germany gave the order for unrestricted submarine warfare to commence on 1 February. An all-out offensive was duly launched by U-boats against Allied shipping in the First World War. For the Allies, locating these unseen killers, which travelled at 8 knots, was a real challenge. The answer lay in the development of underwater listening devices.

## Interim solutions

The first underwater detectors were so-called 'passive' listening devices – simple hydrophones that picked up the sound of a vessel's engines or of its hull moving through the water. Early

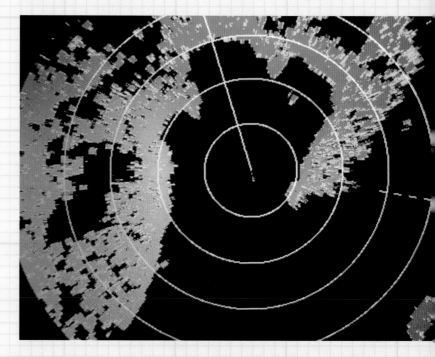

**Keeping watch**
*An active sonar screen on board an oceanographic survey vessel off the coast of New Zealand (above).*

**'Golden ears'**
*A sonar operator on an American submarine in the Second World War (left). Operators like him were selected for the task from the submarine crewmen and gained their nickname for their ability to hear enemy vessels.*

active echo sounders were designed by the Norwegian Einar Berggraf in 1905 and an English meteorologist, Lewis Richardson, soon after the *Titanic* disaster in 1912, but these were never built. The principle involved emitting a sound wave and analysing its echo to determine ocean depth or distance from obstacles like icebergs, reefs or other ships.

## The role of ultrasound

The foundations of sonar – the acronym stands for 'SOund NAvigation and Ranging' – were laid in 1915 by the French physicist Paul Langevin and Russian electrical engineer Constantin Chilowski. Working with British scientists, their research concentrated on the use of quartz crystals and pulses of ultrasound. It was a Canadian physicist, Robert Boyle, who produced a prototype for testing sonar in mid-1917, but it was too late to play a major role against U-boats in the First World War.

In 1943 sonar played a key role in securing the Allied victory in the Battle of the Atlantic, enabling surface destroyers to locate U-boats and depth-charge them. The post-war boom in electronics saw active sonar systems grow more compact, and they have now found a wide range of applications, including detecting shoals of fish, locating wrecks, mapping the ocean floor or in undersea oil exploration.

## PIEZOELECTRICITY

Discovered by Pierre and Jacques Curie in 1880, the piezoelectric effect is the ability of certain materials, such as quartz, to produce electricity when subjected to mechanical stress. Conversely, they change shape when an electric field is applied. Paul Langevin and Constantin Chilowski were the first to built an ultrasound device utilising this effect.

# BCG vaccination 1906

The early years of the 20th century were marked by a tuberculosis pandemic, which was especially prevalent among poorer and disadvantaged groups in society. The bacillus responsible for causing TB had been isolated by the German bacteriologist Robert Koch in 1882, but no cure had been found. Efforts to combat the disease were limited to advising people on how not to catch it and isolating anyone who did contract it in sanatoria for

**Death stalks the streets**
*A French poster of 1918 warns of the 'great scourge' of tuberculosis. A widespread problem in the 19th century, TB made a deadly comeback during the First World War.*

---

### THE LÜBECK DISASTER

In 1930, in the German city of Lübeck, 256 newborn babies were inoculated with the BCG vaccine. More than 130 of them subsequently contracted tuberculosis and 71 died. It transpired that the vaccine had been accidentally contaminated in the laboratory where it was made. The ensuing court case exonerated BCG and its discoverers of any blame, but for a long time its reputation was tarnished by this tragedy.

---

rest cures and plenty of fresh air. Two French scientists were about to change all this: Albert Calmette, a doctor and director of the Pasteur Institute which opened in Lille in 1895, and his colleague Camille Guérin, a veterinary surgeon who joined the institute in 1897.

## Harmless bacilli

The first breakthrough came in 1906, when Guérin showed that resistance to tuberculosis was associated with the presence of living but weak bacilli within an organism. This persuaded him and Calmette that they might be able to develop a vaccine by lessening the virulence of the tuberculosis bacillus. Vaccination with these harmless bacteria would, they hoped, stimulate the immune system against TB without putting people at risk of contracting the disease.

In 1908 they began a series of trials with bovine tuberculosis bacilli, which they subcultured every three weeks on slices of potato cooked in beef bile with glycerol.

**Early vaccination**
*In the early days, vaccinations against TB were performed using a cylinder-like multipuncture unit. Forty small needles were pressed through a piece of absorbent paper steeped in the vaccine and into the patient's arm. Intradermal injections are the preferred method nowadays.*

The trials soon showed that the bacilli became progressively less virulent, but it was only after 230 successive cultures over a period of 13 years that they finally developed a stable strain.

## The war against TB

The first human vaccination using the stable strain was administered in Paris in 1921 by a paediatrician, Benjamin Weill-Hallé, to a newborn baby whose mother and grandmother had contracted pulmonary tuberculosis. The procedure was a great success and over the following years BCG ('Bacillus Calmette-Guérin') vaccination saved millions of lives.

Today, the threat of tuberculosis has receded in most of the developed world and the BCG vaccination is no longer the norm. But the battle against TB is far from over in developing countries, where despite the widespread use of antibiotics and vaccination, tuberculosis remains one of the most lethal contagious diseases.

# FREEZE-DRYING – 1906
# Improving food conservation

Originally developed to meet the needs of biologists, the technique of conservation called freeze-drying – technically known as 'lyophilisation' – was for a long time confined to research labs specialising in pharmaceuticals. It was only after the First World War that it was applied on a large scale to the preservation of everyday foodstuffs.

In 1811 the Scottish mathematician and physicist John Leslie was the first person to change water vapour into ice directly – that is, without passing through a liquid stage in between. Two years later the chemist William Wollaston demonstrated the same process to the Royal Society in London. It did not occur to either of them that this effect might one day be used for the preservation of food.

Almost a century later, on 22 October, 1906, Arsène d'Arsonval presented a research paper to the Academy of Sciences in Paris entitled 'On distillation and desiccation in a vacuum by means of low temperatures'. Written with the help of an assistant, F Bordas, in the biophysics laboratory that D'Arsonval ran, the paper addressed a subject that was preoccupying many biologists at the time: how to preserve samples of tissue, serum, germs and other substances used in research. D'Arsonval had devised a technique of freeze-drying. Three years later an American, Leon Shackell, rediscovered the process and developed it.

## Stages in the process

Freeze-drying involves three stages. First, the sample is frozen solid. Second, it is subjected to

---

**INCA KNOW-HOW**

Father José d'Acosta, a Spanish Jesuit priest on a mission in Peru, reported in 1591 that the Incas conserved food crops by carrying them up into the Andes above Machu Picchu. There, the combined effects of the cold, the Sun and the relatively low atmospheric pressure at high altitude caused the water in the food to sublimate. The resulting products, which the Incas called *charqui* (dried meat) or *chuno* (dried potatoes), were the earliest known freeze-dried foods.

---

sublimation – a process that transforms a solid into a vapour without going through a liquid phase. By this stage, the sample has lost most of its water through evaporation, with no heat involved. Finally, to rid the sample of any residual water, it undergoes a secondary desiccation process, in which the temperature is raised slightly, although still usually staying below zero. The end product is almost totally dehydrated but retains structure and colour.

*Looking to the future*
*Arsène d'Arsonval (below) in 1910, photographed in a newly established laboratory at Nogent-sur-Marne which he headed until 1931. D'Arsonval once remarked that 'science makes yesterday's impossibility tomorrow's commonplace'. He did not patent his freeze-drying process, which was of enormous benefit to biologists, medical patients and cooks alike.*

---

**PROS AND CONS**

Most micro-organisms need water to survive and grow; freeze-drying is very effective at preventing the growth of microbes and inhibiting harmful chemical reactions. The use of freeze-dried products in the Second World War demonstrated the huge benefits they offer in storage and distribution. If properly freeze-dried, most foods, either raw or cooked, have a long shelf life even at room temperature and can be reconstituted simply by adding water; the resulting products provide almost the same nutritional value as fresh versions. The main disadvantage of freeze-drying is its relatively high cost, which has hampered the full development of the technique even to the present day. Also, while freeze-drying commodities such as coffee and tea, which have no cell structure, is successful and cost-effective, other products do not fare so well. It is especially hard to sublimate all the water vapour from foods with abundant cell membranes like meat, vegetables and fruit.

**Medical application**
*Scientists monitor the freeze-drying of an antibiotic in a laboratory (right). Samples are sealed into airtight containers under very high pressure to create the vacuum required for sublimation.*

CONTROLER LE TAUX D'OXYGENE AVANT DE

*nouveau* POTAGE MAGGI

CRÈME DE VEAU

**Convenience foods**
*A 1956 poster for veal soup, freeze-dried by the Swiss food manufacturer Maggi.*

Crucially, many of the biological functions recover once the sample is rehydrated, since freeze-drying does not damage the molecular structure or cell tissue. Neither ordinary drying nor freezing alone produces such a satisfactory result.

## Impact of the war

Because of its cost and complexity, freeze-drying was long restricted to scientific laboratories. This changed in the 1930s. With another world war looming, there was an urgent need to be able to stockpile large, transportable quantities of blood plasma for transfusions to casualties. From 1935, the American Earl W Flosdorf published the results of his efforts to freeze-dry human blood serum and plasma for clinical use. The desiccation of blood plasma from a frozen state, performed by the American Red Cross for the US armed forces, was the first extensive use of freeze-drying. Flosdorf, together with researchers in Britain under Ronald Greaves, pioneered large-scale commercial freeze-drying of foodstuffs at this time. Meanwhile, Ernst Boris Chain, the co-discoverer of penicillin, initiated the lyophilisation of antibiotics and other sensitive biochemical products.

## From drugs to freeze-dried coffee

At the end of the war, the pharmaceutical and cosmetics industries adopted freeze-drying for vaccines, drugs and other preparations. Freeze-dried coffee, which had been brought to Europe by American GIs, helped to stimulate the freeze-drying of foods. At first, because the process was still very costly, it was only used for luxury items. But before long freeze-dried soups, spices and even entire prepared meals appeared in packets on grocers' shelves. In the 1960s NASA adopted freeze-dried meals for astronauts on its space programme.

**Vacuum packed**
*Freeze-dried food was ideal for astronauts. This spread was served up on board the International Space Station sent into Earth orbit in 1998.*

# A new way of building

The contours of the Maison Hennebique, at Bourg-la-Reine in the suburbs of Paris, speak volumes about the material it is built from: reinforced concrete. By creating a plethora of different sculptural and decorative features, the architect François Hennebique made a bold statement to his contemporaries about the amazing possibilities of this new material.

Until the mid-19th century, concrete was used as mortar between blocks of building stone. It was a gardener at the Palace of Versailles, Joseph Monier, who first had the idea of grinding it into a powder and reinforcing it with iron bars; the plant tubs he produced were durable and cheap. His compatriot François Hennebique had far grander plans for reinforced concrete.

## First industrial buildings

Trained as a stonemason, Hennebique (1841–1921) became interested in reinforced concrete in the 1870s. He made flooring from it – its first recorded use in the building trade – in 1879. Then, in 1892, he patented a system of steel-reinforced beams, slabs and columns, an invention that helped to secure the financial future of his modest construction firm. Two years later, his 'ferro-concrete' system was used to build a railway bridge at Viggen, Switzerland.

Hennebique went on to construct a range of industrial buildings using the same method, including coal bunkers, grain silos, the Saint-Ouen oil refinery in Paris (1894) and a

**Stone ship**
*The launch of the Cretell (right), a coal barge with a concrete hull, at Tilbury Docks on 1 May, 1919.*

### FLOATING HULK

French inventor Joseph-Louis Lambot was the first person to make a small boat from ferro-cement. It was exhibited at the 1855 World's Fair in Paris. Yet the idea of concrete vessels went back to ancient times. Archimedes' principle – that all bodies immersed in water are buoyed up by a force equal to that of the weight of the displaced liquid – holds good whatever the material. A concrete ship, if suitably shaped, will float. Steel shortages after the First World War prompted the British Admiralty to commission concrete coal barges. At the same time, to try to minimise the danger of sparks from metal hulls, American shipbuilders designed concrete-hulled oil tankers up to 135m in length.

**Pride of Liverpool**
*The Royal Liver Building (left), built on the Pier Head at Liverpool in 1909–10 by Thomas Audrey, was one of the first high-rise buildings constructed from reinforced concrete. It rises to a height of 90m.*

### NOT-SO FIRE RESISTANT

Hennebique believed that reinforced concrete would help to prevent buildings collapsing in a fire. In fact, the opposite proved to be true: expansion of the steel framework under extreme heat caused the rods to bend and twist so violently that the concrete sheathing exploded. Of traditional building materials, plaster is one of the best fire retardants.

six-storey mill in Swansea for the millers and corn merchants Weaver & Co. This last, built in 1897, was the first reinforced concrete building in Britain and it proved its strength in February 1941 when the city was blitzed by the Luftwaffe for three consecutive nights.

Soon, architects and engineers were beating a path to Hennebique's door. In 1898 he set up a structural engineering consultancy that specialised in the material and he founded the technical journal *Le Béton Armé* ('armoured concrete') to promote it. He also established a network of agencies worldwide. At the 1900 Paris Universal Exposition, Hennebique's cantilevered concrete staircases graced the Petit Palais, confirming him as the world's foremost reinforced concrete contractor. By

**Sweeping lines**
*In 1916–24, Eugène Freyssinet was commissioned to build a series of enormous airship hangars at the Orly airfield outside Paris. These soaring parabolic structures – sadly destroyed at the end of the Second World War – were constructed from 40 separate arched sections, each 7.5m thick and with a span of 86m. When fitted together they created an enormous enclosed space. The arches were cast in situ using moveable formwork.*

**Prefabricated walls**
*These concrete slabs (below) display a double layer of steel reinforcement, designed for use in load-bearing surfaces.*

1909, Hennebique and his franchises around the world had completed 20,000 commissions. Only the First World War halted his rise.

The introduction of reinforced concrete to Britain owed much to the French engineer Louis Gustave Mouchel, with whom Hennebique worked closely during the design and construction of the Swansea Flour Mill. With his business expanding across Europe, Hennebique offered Mouchel an exclusive license to use the technique in the UK; it was Mouchel who coined the term 'ferro-concrete'. Mouchel's first commission, in 1897, was to construct a retaining bank for the London & South Western Railway at Southampton.

## The rise of prestressed concrete

Hennebique died in 1921, just as a new method of pre-stressed concrete construction was emerging. In 1912 a young French civil engineer called Eugène Freyssinet (1879–1962) had learned the limits of reinforced concrete when the truss spans of the Pont la Veurdre, a concrete bridge he had built over the River Allier near Vichy the previous year, began to twist out of shape. Freyssinet's answer was to try to enhance the elastic qualities of concrete by compressing it. By adding pre-stressing tendons of high-tensile steel cable during manufacture, the concrete was made more uniform in composition and therefore less liable to deformation ('creep') and fracture.

### MAKING REINFORCED CONCRETE

Although concrete is highly resistant to crushing, traction forces can soon cause it to break. When a beam is set on two supports and a weight placed on top of it, the beam is subject to two different forces: the upper surface is compressed, while the lower surface bends and stretches. This traction may be powerful enough to snap the beam. In reinforced concrete, steel rods are used to contain this force. Manufacture of reinforced concrete begins in precisely the same way as ordinary concrete: cement, aggregates, sand and water are mixed, along with various other additives designed to enhance certain features (such as shortening the drying time). The metal framework is then sunk into the item. The grade of steel used and exact configuration of the frame depends on the loads that the particular section will have to withstand.

## AN INDISPENSABLE TOOL

Originally labourers prepared cement by tipping the ingredients onto the ground and mixing them by hand with shovels and trowels. But the demand for ever-larger quantities of cement prompted the invention of cement mixers, which churned everything up in a rotating drum. Vanes inside the drum ensured a thorough mix. The first mixers were cone-shaped hoppers that sat upright and were fitted with simple iron crossbraces; as the cement, sand, aggregate and water were tipped in, the braces helped to blend it all together. Then all that was needed was to open a small hatch at the bottom of the hopper and let the mixed concrete pour out. From 1857 onwards, the French engineer Ernest Cézanne began using steam power to drive the drum faster and produce a better mix. The now-familiar lorries that deliver ready-mixed cement to construction sites date from the 1940s.

A German engineer, C Doehring, had described the principle of reinforcing precast concrete with tensioned wires in 1886, but Freyssinet was the first to put it into practice. Freyssinet published his findings in 1926 and three years later left his job with the Claude Limousin construction firm to go into business on his own, manufacturing pre-stressed concrete electricity pylons. He introduced a number of innovations, such as vibrating the freshly poured concrete to improve settling and drying

it artificially to speed up the hardening process. Yet for all his efficiency (and he could turn out 50 pylons a day), the Depression brought a slump in orders and Freyssinet's business folded. In 1933-5, he used his technique to shore up the foundations of the maritime railway station at Le Havre, which was in danger of subsiding. This sealed his reputation and kept his order book full.

## New freedom

Unshackled from the constraints imposed by traditional materials, architects began to design buildings with all the freedom of sculptors. The Catalan architect Antoní Gaudí began using concrete from around the turn of the 20th century, even before the advent of pre-stressing. It proved the ideal material for his highly irregular, fluid forms, as seen on the façade of the apartment block in central Barcelona known as *La Pedrera* ('the quarry'), which he built between 1906 and 1912.

After Gaudí, reinforced concrete inspired some of the foremost architects of the 20th century. In 1948 the Italian Pier Luigi Nervi created a magnificent hemispherical ribbed canopy for the Exhibition Hall in Turin, which was remarkable both for its aesthetic impact and its technical bravura. Concrete was used extensively in capital projects to rebuild cities flattened by bombing in the Second World War. Le Corbusier used it to build his 'radiant cities' (groups of housing units) in Marseilles and Nantes. In both Europe and the USA, concrete rivalled steel as the material of choice for skyscrapers and in 1953 led British architects Alison and Peter Smithson to coin the term 'Brutalism' from the French *béton brut* or 'raw concrete', which Corbusier himself used to describe his constructions.

**Work in progress**
*The enormous and imposing Sagrada Familia in Barcelona (below) was begun by architect Antoní Gaudí in 1883 and is still under construction. The design for the cathedral incorporates a mixture of sandstone, granite, porphyry and reinforced concrete.*

**Bold statement** *UNESCO's headquarters in Paris (above) was built in 1954–8 to plans by Marcel Breuer, Pier Luigi Nervi and Bernard Zehrfuss. The seven-storey, 'Y'-shaped building is supported by 72 reinforced-concrete stilts ('pilotis').*

In the 1960s the cheapness of concrete led policymakers and planners to undertake massive slum-clearance and erect high-density, high-rise housing. The most notorious of these projects in Britain included the Park Hill Estate in Sheffield, built in 1961, and Robin Hood Gardens in Poplar, East London, completed in the 1970s. In the same period the rise of the package holiday encouraged rapid construction of high-rise hotels on unspoilt stretches of Mediterranean coast. Such insensitive development gave concrete a bad name, as 'concrete jungle' became shorthand for the ills of urban living. Even so, at the start of the 21st century, concrete is unchallenged as the world's favourite building material, with more than 9 billion cubic metres used every year.

**Modern masterpiece**
*The Exhibition Hall in Turin (above) was designed by Pier Luigi Nervi and built in 1948. Its most striking feature is its undulating concrete canopy.*

**Radical departure**
*Swiss architect Le Corbusier used innovative construction techniques in his chapel at Notre-Dame-du-Haut (1955). The roof is a hollow shell made from two rough-cast concrete membranes, resting on thick walls of whitewashed local stone.*

69

# Magic of the moving image

The cartoon was once looked upon by the intellectual élite as a minor film genre for children. In truth, it is a fully fledged art form in its own right, characterised by technical innovation and creative virtuosity. It goes far beyond the hand-drawn image, embracing such diverse techniques as stop-motion animation, claymation and CGI.

Visitors to the Grévin Museum in Paris on 28 October, 1892, were confronted by a truly extraordinary sight: drawings projected onto a screen and moving before their very eyes. To their amazement, they watched a young girl offer a bunch of flowers to a surprised Pierrot, while Harlequin lurked in the shadows. These flickering animated sketches are reckoned to be the first examples of projected cartoons. They were created by the French inventor Émile Reynaud and were a spectacular success, attracting over half a million visitors to the moving picture show he called the 'Théâtre Optique' or 'Pantomimes Lumineuses' in the years up to the turn of the century.

This was not Émile Reynaud's first invention. A self-taught jack-of-all-trades, in 1877 he had invented the Praxinoscope, a device that used a combination of mirrors and a strip of pictures attached to the inner surface of a spinning cylinder to produce the illusion of movement. His Théâtre Optique arose from the idea of replacing the cylinder with glass plates set in leather bands; the images on these were then projected onto the screen via a system of rollers, magic lanterns and mirrors. As each of the image bands was around 50 metres long, the scenes could last for several minutes.

## The first shorts

In 1906 the Anglo-American producer James Stuart Blackton, founder of Vitagraph Studios, created the first film to use images on a strip of celluloid, entitled 'Humorous Phases of Funny Faces'. Two years later, on 17 August, 1908, the French caricaturist Émile Cohl premièred his 2-minute *Fantasmagorie*, the world's first fully animated film, at the Gymnase Theatre in Paris. It begins with a hand drawing a clown. The hand then vanishes and the clown transforms into a portly gentleman sitting in an auditorium. When a woman sporting a large feathered hat comes

**All done with mirrors**
*Émile Reynaud's Praxinoscope of 1876 (above) used 12 mirrors to reflect the images; this enabled several viewers to use it at the same time.*

### ATTEMPTS TO CONJURE UP MOTION

The 19th century witnessed a proliferation of animation devices with by exotic names such as the Thaumatrope, Phenakistoscope, Praxinoscope and Zoetrope. The Phenakistoscope, invented by the Belgian Joseph Plateau in 1832, consisted of two discs mounted on a single axis. One disc carried different images of an object that were viewed through slits in the other disc; rotating the discs produced an illusion of motion. British printer John Barnes Linnett patented a flip book, or 'kineograph', in 1868. This consisted of a cartoon strip cut up into individual frames and bound into a book; when the viewer riffled through the pages, the images appeared to move. In 1824 Dr John Ayrton Paris came up with the Thaumatrope – two cardboard circles glued back-to-back, with a string threaded through the centre. When the string was twirled, the pictures on either side of the circle – say, flowers on one side and an empty vase on the other – merged into a single image. All these devices relied on the phenomenon known as 'persistence of vision'.

**Phantasmagoria**
*The image disc of a Phenakistoscope from 1833 (far left). The moving images produced by the device were called 'phantasmagoria'.*

## A PROLIFIC TALENT

Émile Cohl was born Émile Courtet in Paris on 4 January, 1857. He changed his name at the age of 22. As a child, he was obsessed by drawing, a talent that he would use to many different ends. He produced political caricatures for newspapers, sketched costumes for light opera and even designed stamps. Other occupations included performing as a magician's assistant, writing and devising word games, puzzles and riddles. He was a member of several artistic circles in Paris, such as the 'Hydropathes', who specialised in shocking the bourgeoisie, and the 'Incohérents', whose exploits made them forerunners of Dadaism and Surrealism. At the age of 51, Cohl had the brainwave of combining drawing and film, which resulted in his short animation *Fantasmagorie* (right). He worked for various studios in France and the USA, including Gaumont, Pathé and Éclair, but sadly fell from favour and died penniless in obscurity in January 1938.

**Lucky mascot**
*Otto Messmer's Felix the Cat became hugely popular in the 1920s, but his star waned when Mickey Mouse appeared. His associations with good fortune (his name means 'lucky' in Latin) saw him adopted as a mascot by both Charles Lindbergh and some US squadrons in the Second World War.*

**Disney icon**
*Mickey Mouse's first appearance was as a frustrated aviator trying to emulate the exploits of Charles Lindbergh in the silent short* Plane Crazy *in 1928.*

film at 16 frames per second. Cohl, a prolific and endlessly imaginative animator, went on to make more than 300 short films between 1908 and 1923.

Meanwhile, on the other side of the Atlantic, the cartoonist Winsor McCay was busy turning out his famous newspaper comic strip 'Little Nemo in Slumberland'. His growing interest in animated cartoons led him to make the short film *Gertie the Dinosaur* in 1914. Screenings involved McCay standing on stage and interacting with his cartoon creation by pretending to make her perform tricks.

### Masters of their art

Five years later, also in the USA, Otto Messmer and Pat Sullivan created Felix the Cat, the first genuine star of animated films. Another new era dawned in 1928, when Walt Disney unveiled the first cartoon film with a synchronised soundtrack. Entitled *Steamboat Willie*, its hero was one Mickey Mouse. This curious-looking figure had stick-like legs ending in outsized boots, a white face topped by perfectly circular black ears, with two black splotches for his eyes and another for his nose. He might not sound promising, but he and his creator would go on to international fame and fortune.

Whereas the pioneers of cartoon film had single-handedly drawn literally thousands of images, whole teams of illustrators were now employed by an ever-growing number of studios. Illustrators moved from one to another, spreading their know-how. Often based on comic strips, many of the films made at this time were not especially original, but

and sits in front of him, he plucks out the feathers one by one before transmogrifying back into the clown and going off on a series of adventures.

Like Blackton's film, Cohl's drawings were white-on-black and moved extremely fast. What distinguished this groundbreaking film from Émile Reynaud's invention was a simple but revolutionary rule-of-thumb: a separate drawing was produced for each and every phase of the action before being committed to celluloid. To run his film, Cohl used the Cinematograph projector invented some years before by Auguste and Louis Lumière, which set the industry standard of 35mm

**Heigh Ho!**
*An image from Walt Disney's* Snow White and the Seven Dwarves. *Disney used the dancer Marge Champion as a model to make the movements of his heroine more realistic.*

**Walt's war effort**
*Walt Disney (holding the pointer) briefs scriptwriters and US Treasury officials on the storyboard of the 1941 animated short* The New Spirit. *Starring Donald Duck, the film was made to encourage people to pay their taxes promptly to help the war effort.*

---

### THE ILLUSION OF MOVEMENT

Human visual perception is far from perfect: rather than disappearing instantly, an image captured by the retina persists for a fraction of a second after the image is no longer there. So if another image instantly follows on from the first, it becomes superimposed on the latter, so creating an impression of movement. This physiological defect effectively forms the basis of animated film, which runs a succession of static images at such a speed that the eye creates the illusion of movement. While early animation devices utilised this principle of retinal persistence, it was only with the advent of the Lumière Brothers' Cinematograph, which combined photography, animation and projection, that inanimate objects ranging from drawings and puppets to models made from clay really came to life.

---

they did help illustrators to hone their techniques (one time-saving ploy was to keep the same background over several frames).

In the Disney studios, meticulous attention to technical detail and streamlined organisation paid dividends. Short films came thick and fast, as Mickey Mouse was joined by Pluto, Goofy and Donald Duck. The intervening years also saw the introduction of colour film. The time was ripe for the first full-length animated film, *Snow White and the Seven Dwarves*, which was released in 1937 to great critical acclaim. From here on, highly skilled teams of illustrators, directors, animators, colorists and set designers worked in Disney's model studios developing and honing a distinctive Disney style.

## Animation's golden age

Yet Disney was just one of many animation studios that flourished in this period. In 1930, for example, the Fleischer Brothers introduced two cartoon stalwarts: the coquettish nightclub singer Betty Boop and Popeye the Sailorman. In 1942 they adapted the comic strip Superman (created in 1938 by Jerry Siegel and Joe Shuster). From 1938 onwards such firm favourites as Bugs Bunny, Tom and Jerry, Tweetie Pie, Sylvester the Cat, Road Runner, Wile E Coyote and Speedy Gonzales all made it to the big screen. In Europe cartoon subjects tended to be more highbrow, including classics such as Berthold Bartosch's *The Idea* (1930) and Alexandre Alexeieff and Claire Parker's *A Night on Bare Mountain* (1933).

The unstoppable rise of cartoons continued after the Second World War. In Canada, Norman McLaren released *Blinkity Blank*, which won the short film prize at the 1955 Cannes Film festival. McLaren's technique was utterly unique, in that he scratched the images directly onto the surface of the film.

The stop-motion technique was first used in film-making as early as 1898 by Englishmen Albert E Smith and J Stuart Blackton, who brought a toy circus of acrobats and animals to life in *The Humpty Dumpty Circus*. The first woman animator, the American Helena Smith Dayton, began experimenting with clay stop motion 1916 and in the same year released her first film, *Romeo and Juliet*. The Czechoslovak filmmaker Jiri Trnka used puppet animation in a similar way in his retelling of familiar folktales in the 1940s.

## CARTOON AS PROPAGANDA

The film-maker John Halas and his wife Joy Batchelor began making short animated propaganda films during the Second World War. In 1954 they created *Animal Farm*, the first British full-length animated feature for adults on general release. Quite unknown to them, their adaptation of George Orwell's allegory on a totalitarian state was secretly funded by the CIA, which at the time was keen to promote any art with an anti-communist message.

## LABOUR-INTENSIVE

A 15-minute cartoon involves 45,000 separate images, or 3,000 images for every single minute of film.

### Betty Boop
*The Hayes Code, which was in force in the USA from 1934 to 1966, forced the creators of this flirty cartoon character to lower her hemline to make her less sexy.*

### Towering achievement
*The inventive French cartoon* The King and the Mockingbird *(1980) is considered by some to be the best animated feature film of all time.*

In Japan animation was pioneered by Noburô Ôfuji, who specialised in cut-out paper silhouettes. One of Ôfuji's most famous works was *The Whale* made in 1927. He was followed by Osamu Tezuka, a comic-strip artist who moved over to film animation and is regarded by many as the founder of 'Manga' illustration. The most renowned Japanese animator today is Hayao Miyazaki, creator of acclaimed full-length features such as *My Neighbour Totoro* (1988) and *Ponyo* (2008).

The rise of television in the 1960s opened up a new market for cartoons. The most successful have been syndicated American series, such as Hanna-Barbera's *The Flintstones* and *Yogi Bear* and,

more recently, *The Simpsons* by Matt Groening and the outrageous adult show *Family Guy* by Seth MacFarlane and David Zuckerman. Many of these series spawned spin-off merchandising.

**Japanese genius**
*Spirited Away (above), a full-length film by Hayao Miyazaki and his company Studio Ghibli, won many awards including the Oscar for Best Animated Feature in 2003.*

**Homer's odyssey**
*Since first airing on TV screens in 1989, The Simpsons cartoons by animator Matt Groening have enjoyed extraordinary success, transferring to cinema screens in a full-length feature in 2007. The dysfunctional family of Homer, Marge, Bart, Lisa and baby Maggie offer a wry take on modern American life.*

## FRAME BY FRAME

A technique employed in the very earliest days of animation, and recently enjoying a revival, is stop-motion. This involves taking individually photographed frames of an object that is moved slightly, by hand, between frames. When the frames are run together, the object appears to move. Czech film-maker Jiri Trnka used stop-motion to animate jointed marionettes. Variants include cut-out silhouettes, as employed by Noburô Ôfuji, and even sand (as in Co Hoedeman's *The Sand Castle*, 1977). More recently the technique known as claymation, which uses malleable figures in modelling clay, has been used to great effect by Nick Park in the Wallace and Gromit films, Tim Burton in *The Nightmare before Christmas* (1993) and Wes Anderson in *Fantastic Mr Fox* (2009), where it is combined with other types of stop-motion.

## The rise of CGI

Traditional cartoon films were produced by a process known as cel (short for 'celluloid') animation. Illustrators created characters on transparent cellulose acetates that were then used as overlays on a static background drawing. This was time-consuming, labour-intensive work: Disney's *Snow White and the Seven Dwarves*, for example, was three years in the making. By the early 1990s, a new approach had arrived in the form of computer-generated imagery (CGI), in which images are composed of tiny digital units called pixels. Used initially to create two-dimensional images, but with growing computer power increasingly applied to 3-D, CGI came of age in 1995 with *Toy Story*, the first completely computer-generated film. This was followed by other global box-office hits such as *Shrek* (2001), *Finding Nemo* (2003) and *Up* (2009).

**Back to the future**
*The Gromit half of the duo Wallace and Gromit (above), as he appeared in* The Curse of the Were-Rabbit *in 2005. To create feature films with clay characters, British director Nick Park of Aardman Animation revived the old technique of stop-frame animation.*

**Making history**
*Toy Story (1995), created by Pixar Studios, was the first cartoon feature film in which the images were entirely computer-generated. Two of the movie's main characters (right) were Buzz Lightyear and Woody the Cowboy, whose voices were brought to life by actors Tim Allen and Tom Hanks .*

# Streamlining manufacturing

In 1913 a modern industrial revolution took place when Henry Ford introduced the assembly line to his automobile plant in Detroit. Under this new approach, every worker had a work-station where one specific task in the car manufacturing process was performed, before the car rolled on down the line to the next worker. This radical innovation produced huge time and cost savings and put Henry Ford in the driving seat of car production worldwide.

In the early years of the 20th century, Henry Ford had a dream of making a car for the common man. At that time automobiles were playthings of the rich, hand-built to an individual's specifications. Ford decided to break with this cottage-industry craftsmanship and offer instead a single, high-quality but low-cost model. From 1908 onwards, he began manufacturing his 'Model T', the world's first mass-produced car. Popularly known as the 'Tin Lizzie', it found a ready market, with sales topping 5 million vehicles within 20 years. On the eve of the First World War, one in every two new cars on American roads was a Model T Ford.

## Putting Taylorism into practice

Ford achieved this remarkable success by the systematic application of the principles of 'scientific management' devised in the late 19th century by the American engineer Frederick Winslow Taylor (1856–1915). At Highland Park, a suburb of Detroit, Ford built a brand-new factory in which production efficiency was maximised by standardisation. Identical interchangeable parts meant that there was no need to tweak components by hammering, filing or cutting them down prior to assembly. Orders came thick and fast: from sales of 18,664 cars in 1909-10, output shot up the following year to 78,440 cars. This success put a premium on producing even more units at an ever-faster rate.

## Assembly repetition

Having achieved what he could by efficiency and standardisation, Ford needed a more radical approach to meet the soaring demand. At the time, factories were divided into separate workshops and each automobile would be sent from one workshop to the next to have, say, the engine, the axles and wheels, or the interior fittings installed. Once in the appropriate workshop, the car sat in its spot

### HENRY FORD

Henry Ford was born in Dearborn, Michigan, in 1863, the son of an Irish farmer who had emigrated to the USA. As a young man, Henry's burning interest was engineering and he built his own car from scratch. In 1903 he founded the Ford Motor Company, which soon became the leading automobile manufacturer in America. He diversified into making aeroplanes at the outbreak of the First World War. Autocratic and paternalistic, he was vehemently opposed to the interference of trade unions and government alike in private enterprise. Ill-health forced him to hand over the reins of power to his grandson Henry Ford II in 1945. He died two years later, aged 83.

#### Going places
*Ford invented a petrol engine in 1893. In this photograph (right), taken three years later, he is at the wheel of a car he had designed and built himself. It had aluminium bodywork and a four-stroke engine, but was not equipped with brakes or reverse gear.*

while the workers moved around it, taking turns at the different operations involved in the car's assembly. From 1913 Ford turned this arrangement on its head: each worker had a specific work-station and performed a specific task – fitting a screw, say, or a nut, tightening a bolt, or installing a magnet – before passing the component onto their neighbour for the next operation. As Ford put it: 'Save 10 steps a day for 12,000 employees and you will have saved 50 miles of wasted motion.' The upshot of his new process was that by 1914 the number of minutes required per worker to assemble a car had been reduced from 840 to just 93. Output doubled from the previous year, while the number of people employed, which till then had grown steadily year on year, suddenly dropped from 14,336 to 12,880. Ford's gamble had paid off.

Assembly-line manufacture became the norm after the Second World War. The work was monotonous, but most employees were motivated by the high wages made possible by the increased productivity. The sustained growth of the post-war period stimulated the emergence of a new middle class among factory employees, but the model reached its limit in the 1970s as consumers began to demand a wider range of goods. A further factor was that a new generation of workers were less happy with the relentless monotony of the assembly line, leading to absenteeism, high turnover and slipshod workmanship. Robots, first introduced to car-making by General Motors in 1961, would soon take over many jobs once carried out by humans.

**'Any colour so long as it's black'**
*In 1916 Henry Ford ordered all his cars to be painted black, as this was the fastest-drying paint colour. The Model T above was built before that date.*

**Standard practice**
*'Fordism', the mass-production of goods on assembly lines, spread around the world after 1945. These women workers at a factory in Berlin are assembling record players in the 1950s (right).*

# Skimming over the waves

**P**eople strolling by Lake Maggiore in Italy one fine day in 1906 were astonished by a strange craft anchored just offshore. It was some 4 metres long and had huge propellers mounted fore and aft on a shaft suspended above its slim hull. The first hydrofoil flight was about to take place.

**Curious craft**
*Enrico Forlanini's hydrofoil moored on Lake Maggiore (above right). In 1877 this versatile inventor had also developed an early helicopter; powered by a steam engine, it flew for 20 seconds – a world first.*

At the controls was the Italian engineer Enrico Forlanini, the inventor of the machine. As he started up its 60-horsepower engine, a cloud of blue smoke and a gutteral roar filled the air, then powered the craft along. Suddenly the hull began to lift out of the water, supported by a system of ladder foils. The run was timed, recording a speed of 36.2 knots (67km/h); Forlanini was jubilant.

A simple idea lay behind the invention of this hybrid boat and aeroplane. In a fluid (be it air or water) every body in motion is subject to lift, a force that, depending on the body's shape, will push it either up or down. A wing moving through air was called an aerofoil, and so a wing moving through water came to be known as a 'hydrofoil'; by extension, the name was applied to the whole craft. The idea was first conceived in 1869 by the French engineer Joseph Farcot, but only patented and put into practice by his compatriot Count Charles de

Lambert, who tested a steam-powered design in 1897. The first scholarly paper on the subject, written by an American, William E Meacham, appeared at around the same time. In 1906, the same year as Forlanini's tests, Meacham gave a full account of the principle of the hydrofoil in *Scientific American*.

The great advantage of this new type of craft was that, in flying over the water on its foils, it experienced only a fraction of the drag that slowed down traditional ships and boats. This enabled hydrofoils to attain unheard-of speeds. In 1919 Alexander Graham Bell, the inventor of the telephone, set a record of 61.5 knots (114km/h) in his HD4 hydrofoil, a craft built to his design and fitted with two powerful 350hp engines.

## A concept comes of age

In the 1930s the German naval architect Hanns von Schertel filed several patents for

### SWAMP SKIMMERS

**L**ike early hydrofoils, the distinctive flat-bottomed craft known as airboats are also powered by an aerial propeller. But unlike hydrofoils – or indeed hovercraft, which float on a cushion of air – they do not actually leave the water. The first airboat, the Ugly Duckling, was built in Nova Scotia, Canada, by Alexander Graham Bell. Airboat use grew in the 1920s and 1930s with the development of lighter, more powerful aero engines. Their flat hulls, which have no operating parts below the waterline, make airboats ideal for travelling across shallow swamps and marshland, such as the Florida Everglades (left) and in the bayous of Louisiana. They were also widely used in the Vietnam War.

large passenger-carrying hydrofoils. His ideas eventually came to fruition in Switzerland after the Second World War (at the end of the war Germany was banned from building boats with speeds in excess of 12 knots). In 1952 the Supramar company, founded by Von Schertel, inaugurated the world's first regular hydrofoil service with the 32-passenger *Frecchia d'Oro* ('Golden Arrow'), which plied a route across Lake Maggiore between Italy and Switzerland.

In the late 1960s and early 1970s, the US Navy embarked on its 'Pegasus' programme to build fast, gas-turbine powered hydrofoils to patrol its coastal waters. Six craft were built and have since distinguished themselves, particularly in missions to intercept the fast speedboats used by South American drug smugglers attempting to traffic illegal drugs.

## Speed at all costs

Hydrofoils today are widely used throughout the world as passenger vessels and recently much research has gone into using hydrofoil technology in yacht design. Innovations incorporate new ultralight composite materials and specially shaped hulls and foils that take full advantage of the wind and also minimise cavitation (the formation of vapour bubbles). In 2008 French yachtsman Alain Thébault broke the speed record for sailing boats in his experimental hydrofoil *L'Hydroptère*, clocking up 51.36 knots (104km/h) – close to Bell's achievement in 1919, but this time using only the wind as motive power.

**Flying Frenchman**
*The yacht L'Hydroptère (above), skippered by Alain Thébault, crossed the Channel in 34 minutes and 24 seconds in 2005, beating Louis Blériot's 1909 flight-time by almost three minutes.*

**Quick crossing**
*A modern hydrofoil in service between Greek islands in the Aegean. The hydrofoil is twice as fast as a conventional ferry.*

---

### COUNTERING CAVITATION

Hydrofoil operation can be impaired by a phenomenon called cavitation, which also to some extent affects propellers in conventional ships. As the foils of the craft cut through the water at high speed, air bubbles form on their surface. These bubbles can coalesce behind the foil into a larger bubble with relatively low gas pressure. As the water streams at high pressure around it, the bubble (or 'cavity') collapses, causing a sudden loss of lift, which can also seriously increase drag, and creating a shock wave that can destabilise the vessel. Cavitation is a tricky problem to overcome, and engineers are constantly seeking to streamline the profile of the foils in order to minimise bubble formation.

# Instant remote communication

From the mid-19th century, inventors had been working on ways of transmitting words and images down telegraph or telephone wires. The German physicist Arthur Korn is generally considered to be the inventor of the facsimile, or fax, machine, but it was the Frenchman Édouard Belin who popularised the technology.

History was made on 17 October, 1906, when the German physicist and mathematician Arthur Korn (1870–1945) became the first person to transmit an image across a telegraphic network. Korn sent a photograph of Crown Prince Wilhelm from his laboratory in Berlin to the offices of the French magazine *L'Illustration* in Paris, a distance of some 1,800 kilometres (1,100 miles).

A year later, it was the turn of staff at the *Daily Mirror* to be amazed, when a photo of King Edward VII was wired from Berlin to their newsroom. The newspaper eagerly embraced the new technology and by 1909 was using an improved 'telectograph' system to send pictures from Paris to London and also from Manchester to London.

## CLOCKWORK AND CHEMICALS

The inventor Alexander Bain patented the first device capable of sending a text document via a telegraphic network in 1843. His device used clockwork to synchronise the motion of two pendulums for line-by-line scanning of the original. Three years later he filed another patent, this time for a chemical telegraph that could reproduce a transmitted image on moving paper tape impregnated with a mixture of ammonium nitrate and potassium ferrocyanide. These chemicals produced blue marks when an electric current was passed through the tape. The fact that Bain's device involved fewer mechanical parts made it faster than transmission by Morse code, which by then was commonplace; it could send 282 words a minute compared to 40 by Morse.

**Public demonstration**
*Arthur Korn takes centre stage to demonstrate his new 'Telautograph' system in a photograph published in* L'Illustration *magazine in 1907.*

## A brilliant idea

In 1843 the Scottish clockmaker Alexander Bain patented a rudimentary prototype facsimile transmitter based on a pendulum apparatus. But unlike this and subsequent inventions that built upon it, such as Giovanni Caselli's Pantelegraph of 1863, Korn's Telautograph broke new ground in operating electrically rather than mechanically. Since 1902 Korn had been investigating the photoelectric properties of selenium cells, which were light-sensitive. A photographic negative of the original document was wrapped around a transparent glass cylinder, which rotated; as it did so, the selenium cells 'read' the document by scanning the light and dark areas of the image. These were then converted into electrical impulses and transmitted down a telegraph or telephone line to a synchronised receiver. Selenium cells in the receiver device reconstructed the image on a sheet of paper specially treated with photosensitive chemicals.

Although it was slow and yielded poor-quality results by modern standards, in essence, Korn's invention was the world's first genuine fax machine. It enjoyed some success, being adopted by the police, meteorologists, the military and newspapers. The one major drawback was transmission time: the small photo he sent to Paris in 1906, for example, took around half an hour to arrive.

## Enter the Belino

By the eve of the First World War, Bain and Korn's devices had been superseded by the Belinograph – 'Belino' for short. Named after its inventor, the Frenchman Édouard Belin, it is widely regarded as the true antecedent of fax machines and photocopiers as we know them today. In place of Korn's selenium cells, Belin's machine measured light intensity by means of an electric eye, or photoelectric cell. The document to be transmitted was still rolled onto a cylinder and scanned line by line, but at the receiving end the elements that made up the image were reproduced on photographic paper on a cylinder installed in a darkroom.

The Belino proved more accurate than earlier inventions at reproducing the various greyscale tones in monochrome photographs. It was also faster than its predecessors and rivals, taking just 12 minutes to transmit an image measuring 130 x 180mm. Belin constantly improved his invention, crucially making it portable. It sent the first

### THE PANTELEGRAPH

One of the main problems that bedevilled Bain's prototype fax machine was the near-impossibility of synchronising the transmitter and the receiver. Giovanni Caselli (1815-91), Professor of Physics at the University of Siena, came up with a neat solution in his Pantelegraph of 1863. It involved synchronised pendulums that were regulated by electric pulses from electromagnets at either side of the pendulum's swing. They thus worked independently of the current for the telegraphic apparatus, which was susceptible to atmospheric conditions. A Pantelegraph service operated between London and Liverpool, but was withdrawn following an economic crisis in 1864.

**Photocopy pioneer**
*Édouard Belin receiving a transmission on an early version of his telephoto machine (right). Belin's telegraphoscope receiver of 1907 (below) incorporated an oscillograph, a relay, an equilibrator and an induction coil. Sparks from the latter perforated a piece of paper with tiny holes to form the image.*

**Hot off to press**
*The portable Belino (left) was more compact than a contemporary typewriter, easy to operate and capable of being connected to an ordinary phone line. Not surprisingly, it cornered the press reporting market.*

on computer printers (such as inkjets and laser printers), but it was 1987 before special heat-sensitive fax paper was replaced by plain paper in a machine introduced by Canon. Machines combining printer, fax and photocopier are now commonplace. Transmitting and receiving a fax can be done from an ordinary PC via a standard telephone network.

The fax played a major role in the growth of an IT-literate society. Accepted by courts as legal evidence, it established itself throughout the business world and the civil service as the standard means of sending all forms of documentary proof of transactions and other communications. Though the fax machine itself is now very much yesterday's technology, the concept of transmitting data verbatim lives on in the practice of attaching PDFs or other documents to e-mails.

**Expediting business**
*The fax helped to speed up all kinds of commercial transactions. Here (above), an operator transmits a money order in 1938.*

remote photo news story from the Great War, and in 1933 was used by press agencies to set up the first international photo-wiring service, between London, Berlin and Paris. Belin also improved the process in 1921 so that it was able to transmit images by radio waves.

## Spreading the word

The first public network of what came to be known as the fax (a contraction of the word 'facsimile') was established in the USA by Western Union in the 1930s. Nine years later, the Associated Press news agency set up the first private system. Fax and photocopy quality improved as a result of technologies pioneered

**State of the art**
*Today's modern fax machines, with laser-quality print, produce faxes that are far more legible and durable than old ones made on heat-sensitive paper, which soon faded and became unreadable.*

# Cellophane 1908

**Safety measure**
*When Britain declared war on Germany, on 3 September, 1939, the government advised the public to cover their windows with cellophane. This was to minimise damage caused by glass being shattered in bomb blasts and also to keep out potential poison gas.*

Jacques Edwin Brandenburger (1872-1954) was a Swiss textile engineer who invented cellophane in 1908. He had gained a doctorate in chemistry at the age of 22 and made his discovery while trying to create a durable and waterproof fabric. Cellophane is made by dissolving cellulose in sodium hydroxide and carbon disulphide to create viscose, which then undergoes other processes before emerging as a flexible transparent film.

Brandenburger patented the process in 1917 and, eager to exploit his invention, set up a factory in Paris in 1920. Three years later he sold his patent to DuPont, and the chemical company went on to improve cellophane by making it heat-resistant. It would be used in all kinds of applications, from adhesive tape to food packaging. In medicine, cellophane found a use in kidney machines, contributing to the first successful renal dialysis in 1925.

# The Geiger counter 1908

In 1908 the German physicist Hans Geiger was working in Manchester as an assistant to Ernest Rutherford, the New Zealand-born scientist renowned for his work on radioactivity (Rutherford was awarded the Nobel prize for physics later that same year). Both Rutherford and Geiger were convinced that the emissions of radioactive alpha particles (helium nuclei) were measurable. This led Geiger to develop the prototype of the detector that still bears his name; he perfected his Geiger counter five years later.

In 1928, in collaboration with his fellow German, the physicist Erwin Wilhelm Müller, Hans Geiger introduced a substantially improved version of his device. This new machine could detect all forms of ionising radiation – alpha, beta and gamma particles, as well as X-rays. Its applications were many and varied, ranging from searching for uranium deposits to checking for radioactive leaks from reactors. Nowadays, the Geiger-Müller counter is a vital tool used in detecting and measuring nuclear radiation. Hans Geiger died in 1945.

## A SIMPLE PRINCIPLE

The Geiger counter is a very straightforward but effective device. It consists of a metal cylinder filled with inert gases – usually helium, neon and argon, with halogens added – and running down the centre of the tube is a piece of conductive wire. When radioactive particles enter the cylinder, they ionise the gases. The tube amplifies this conduction by a cascade effect, which passes down the wire and is registered by a needle on a calibrated scale. The device also emits audible clicks that grow in frequency with increasing radiation.

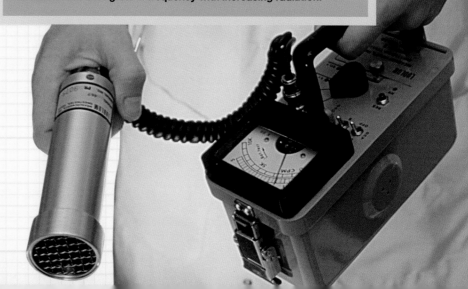

# Movies in colour 1909

**The wonderful world of colour**
*A poster advertising Walt Disney's 1932 production* Flowers and Trees, *part of the 'Silly Symphony' cartoon series, directed by Burton Gillet. The film was notable for being the first to use the new 'three-strip' Technicolor process. Disney outflanked his rivals by striking an exclusive deal with Technicolor.*

On 26 February, 1909, the Palace Theatre in London was host to the world's first public screening of a colour motion picture. The film, entitled *A Visit to the Seaside,* had been made in Brighton by George Albert Smith (1864–1959) and ran for eight minutes.

From the earliest days of cinema, colour had been something of a Holy Grail for directors. Some films had even been hand-coloured, frame by frame, to create the effect. In 1889 the Englishman Edward R Turner had patented a three-colour motion picture system, but it was never sufficiently developed for commercial use. When he died, his American partner Charles Urban sought George Smith's help to take the process further.

Smith was a former music-hall performer – his mind-reading and hypnotism act had been popular back in the 1880s – but he abandoned this earlier career to become a pioneer of the embryonic cinema industry. In 1906 he devised

## GEORGE SMITH'S KINEMACOLOR

Smith's colour process used a filter composed of two colour segments (red and green) and two open segments, mounted on a rotating disc. Prior to shooting, which was still done on black-and-white film, the disc was rotated in front of the shutter and the film exposed successively through the different segments by means of a camera that filmed at 32 frames a second, or twice the normal rate of exposure. The projector worked on the same principle, showing 32 frames a second – 16 each through the green and red filter respectively – in each case with a static image of the opposite colour superimposed over the main colour. The high projection speed and the alternation of the two colours fooled the viewer's eye into perceiving most colours of the spectrum.

a two-colour system that he named Kinemacolor. Private screenings were held in 1908 before its public debut the following year.

**Doomed system**
*A high-speed Kinemacolor projector from 1910. This two-colour process could reproduce most colours (except blue), but the Technicolor system was more reliable.*

## The rise of Technicolor

At first Kinemacolor was very successful, but Smith's enterprise collapsed in 1914 following a patent suit by photographer William Friese Greene. After the First World War it was supplanted by Technicolor, a process that involved cementing strips of different colour together, rather than relying on a high-speed projector. The first two-colour Technicolor movie was *The Toll of the Sea* in 1922, but a three-colour process soon took over.

# A shiny new world

The year 1909 saw the introduction of the world's first completely synthetic plastic, discovered by the chemist Leo Baekeland. Since then a bewildering array of plastics, each with unique properties, has come onto the market. Plastics have become an inescapable feature of modern life, but they are now at the heart of the ecological debate.

In 1889 the Belgian chemist Leo Baekeland and his wife emigrated to the USA, a country they had first visited on their honeymoon. In 1891 Baekeland invented Velox, a photographic paper that could be developed under artificial light, and eight years later the Eastman Kodak company bought the licence to his paper for a million dollars. Unlike many recent arrivals, the couple's financial future was now assured, but Baekeland was not remotely tempted to rest on his laurels. Instead, he continued to conduct research into insulating materials. At the dawn of the 20th century, electricity was the big new thing, and he was keen to play a part in this new venture.

## Resilient and transparent

In the early 1900s, natural resins imported from the Far East were much in demand for use as insulation on the induction coils of transformers. As a result, the price of resins soared. From 1900 onwards, Baekeland directed his efforts towards finding a way of synthesising an artificial resin substitute. His research led him to study the work of the great German chemist Adolf von Baeyer, who in 1872 had tried mixing phenol (a solvent similar to spirit of turpentine) and formalin (a solution of methanol and water). At the time von Baeyer was investigating new methods for manufacturing synthetic dyes and, as far as he was concerned, the hard, transparent residue that his experiments yielded was a failure. In contrast, for Baekeland it suggested some exciting new possibilities.

**New forms**
*A desk lamp of 1945 made from compression-moulded Bakelite.*

**Plastic playthings**
*A range of toys from the 1930s made of Lumarith, a thermoplastic invented in 1912 by the American Celluloid Manufacturing Company. This plastic was based on wood cellulose, a natural polymer.*

## PIONEERING PLASTICS

The long-chain molecules that go to make up plastics can be found in nature – for example in wood cellulose, or in proteins such as keratin in hair and nails, or in the latex exuded by some plants, notably the rubber tree (*Hevea brasiliensis*). The first plastics, called 'semi-synthetics', used such natural ingredients as their starting point. In 1862 the British chemist Alexander Parkes unveiled 'Parkesine' at the international exposition held that year in London. Made from a combination of cellulose, nitric and sulphuric acids and various oils, Parkesine resembled ivory or horn. It was used to make bracelets, bag handles and shoehorns, but the material proved fragile – under certain conditions, it was even explosive. From 1868 on, Parkesine was eclipsed by Celluloid, created by the American Hyatt brothers, who won a competition to find an ivory substitute for the manufacture of billiard balls. A compound of cellulose nitrate and camphor, Celluloid found a wide range of uses, including in knife handles and dolls. It was also used to make the first photographic film. One drawback was its flammability, and it was gradually replaced by other materials in the 20th century.

## Glass orchestra

*In 1946–7, the New York musician Billy Glass bought a stock of Plexiglas and used it to mould almost an entire orchestra of see-through instruments (right). He made them all in his own kitchen by heating the material to 250°C in his oven and using traditional instruments as moulds.*

## The birth of Bakelite

In an attempt to fine-tune the temperature and pressure at the moment when the phenol was brought into contact with the formaldehyde, Baekeland invented the 'Bakelizer', a large, egg-shaped autoclave, which enabled him to heat the ingredients under high pressure. In 1906, after two years of intensive experiments, he finally ascertained that a light heating

## Height of chic

*Bakelite was used extensively by designers and manufacturers in the 1920s, as in this hand-engraved Bakelite powder compact.*

produced a liquid that was ideal for coating surfaces, rather like traditional lacquers. If he increased the heat slightly, the solution became thick and sticky. At still higher temperatures, this paste turned into the hard, colourless material that Baeyer had ended up with. This he called Bakelite and in 1909 he presented it to the American Chemical Society as the world's first entirely synthetic plastic.

Bakelite enjoyed instant success, and not just as an insulator. It was easy to mould and kept its shape even when heated. Its potential applications were almost endless, from aircraft propellers to car dashboards, kitchen utensils, billiard balls, ashtrays and pipes. In 1910 Baekeland founded the General Bakelite Corporation and remained its president until 1939, the year it merged with Union Carbide. He died five years later, aged 81.

## Long-chain molecules

Up to the 1920s, only a trickle of new plastics appeared on the market. But in 1922, the German chemist Hermann Staudinger invented a synthetic rubber he named 'Buna' (from BUtadeine and the chemical symbol NA for sodium). At the same time he explained the theoretical bases for the chemistry of

### POLYMERISATION – THE SECRET OF PLASTICS

All plastics are composed of polymers. These are extremely large molecules made up of the same pattern of atoms multiplied up to 100,000 times. These patterns are mainly found in by-products of petroleum, such as ethylene, propylene or isobutylene. The first people to conceive of polymers, in the early 19th century, were French chemist Henri Braconnot (1780–1855) and Jöns Jacob Berzelius (1779–1848), a Swede, who coined the term 'polymer' in 1830. However, they had no way of identifying them – that development only occurred in the 1920s. It is these huge molecules that give plastics their cohesion. Adding cheap ingredients such as calcium carbonate or silica to the mix enables manufacturers to generate more end-products from a given number of long-chain molecules and also to keep costs down. Such additives alter the density of the finished plastic. Other specific properties, such as colour or resistance to shock, heat or chemicals, are obtained by adding plastifiers (for example, phthalates and phosphates) or colourants.

**Clean and bright**

*A range of new synthetic plastics came onto the market in the late 1930s. Easier to mould than Bakelite, they were produced in all colours of the rainbow (above right).*

### A FAMILY AFFAIR

Plastics are put in three separate categories according to their physical properties:

• **Thermoplastics** take the form of hard granules in their raw state. They are softened by heating, mixed with additives and formed into the desired shape. They can be reheated, reshaped and recycled. Plastics in this family include PVC, polystyrene and polyamides.

• **Heat-hardening plastics** have to be polymerised in a mould into their finished form, since once hardened under heat they cannot be reshaped. Bakelite and the epoxy resin glue Araldite® are in this group.

• **Elastomers**, as their name suggests, are highly malleable. They include synthetic rubbers and Neoprene.

**Lovely legs**

*DuPont's wonder fabric nylon was used not just for stockings, but also to make parachutes for American airborne troops in the Second World War.*

polymerisation. Staudinger demonstrated that plastics are not composed of interlinked rings of molecules, as had once been thought, but rather from long straight chains of thousands of small molecules called monomers. It took some time for the scientific community to accept his thesis, with the result that research into plastics only really took off in the 1930s. Further impetus came from the falling price of oil, which provided the raw material for many of the polymers used in plastics. PVC, Plexiglas and polythene (which Tupperware made into the world's commonest plastic) all appeared before the Second World War.

In this same period Wallace Carothers, a brilliant young researcher at the laboratories of the American chemical giant DuPont, emerged as a leading authority on polymerisation. In 1931 Carothers and his team created Neoprene, which was resistant to heat, light and most solvents. In 1937, the year of his untimely death, he invented a fabric that even after stretching was still stronger than silk. Called nylon, it was introduced the following year in the manufacture of ladies' stockings and toothbrush bristles.

### Plastic planet

The boost given to materials research by the Second World War resulted in the introduction of many new plastics in the 1950s. German

**One small step**
*When Neil Armstrong and Buzz Aldrin made the first moon landing on 21 July, 1969, they were wearing helmets of polycarbonate, a material now used, among other things, to make motorbike helmets and bullet-proof screens.*

**Home help**
*The introduction of Formica, a heat- and chemical-resistant plastic coating that was easy to clean, began to transform kitchen design in the 1950s.*

chemists at the Bayer company developed polycarbonate, a transparent and highly durable polymer that was used to make the helmets for the astronauts on the Apollo 11 space mission of July 1969, which landed men on the Moon. Meanwhile, in 1953 DuPont had created high-density polyethylene (HDPE), an extremely strong material with a high melting-point, which opened up new uses for plastics. As well as bathtubs for babies, for example, it was now possible to make buckets that could withstand boiling water and receptacles for corrosive chemicals. Another new material was Formica, a wipe-clean laminate of paper and melamine resin, which in the 1960s was widely used for furniture.

Plastics came to epitomise the affluent consumerism of the era, giving rise to the idea of the throwaway society, which before long was a reality. Disposable razors were binned after a week, broken appliances and toys were discarded rather than repaired, and plastic bags thrown away as soon as they had been used to carry goods home. Over time it became clear that this was bad for the environment. Most plastics degraded only very slowly, if at all. Current research is concentrating on developing biodegradable plastics, commonly using starch, which deteriorates quickly when subjected to sunlight. One problem, though, is that these new polymers are not as durable as traditional plastics, which still accounted for 260 million tonnes of world output in 2007.

---

### PLASTICS ON THE ROAD

As early as 1942, Henry Ford unveiled a car with plastic bodywork to the press. Picking up an axe, he hit the vehicle's boot to demonstrate how sturdy it was. His experimental vehicle never went into production, but it did highlight the longstanding cooperation between the plastics and automobile industries. As early as the 1930s, dashboards and wing mirrors were being made of Bakelite. The first plastic wing panels, made from heat-hardened polymers, appeared in the 1950s and 1960s. In 1972 the Renault 5 incorporated revolutionary moulded plastic bumpers. Nowadays, plastics are used extensively for car seats and upholstery, internal door and steering-wheel trim, handles, carpets, knobs and switches, petrol tanks and headlight shells. It is estimated that the average car today contains 150kg of plastic (replacing 350–400kg of traditional materials). A prime reason for the increasing use of plastics in cars is to reduce the weight and thereby lessen fuel consumption.

**Environmental hazard**
*In November 2004, strong winds scattered tonnes of plastic debris from a landfill site in the south of France over the surrounding countryside.*

## RECYCLING PLASTICS

Plastics have a long life – they can survive for decades in the natural environment. As they gradually break down into ever smaller fragments, these are ingested by every link in the food chain, with consequences for wildlife that are still only poorly understood. When dumped en masse in landfill sites, they can produce pockets of methane, a major contributor to greenhouse gases. Unregulated incineration of plastics releases hydrogen chloride gas, which can cause acid rain. So all in all, plastics are a major ecological headache. Since the 1990s, the focus has been on limiting the use of plastics, for instance by phasing out the practice of handing out free carrier bags in supermarkets, but tests have also shown that controlled burning of plastic waste in incinerators can yield valuable energy. One plastic bag, for instance, is capable of powering a 60-watt light bulb for ten minutes. Recycling facilities have therefore been set up to reclaim plastics. First, plastic waste must be sorted into the different types of plastic before it can be crushed, washed, dried and ultimately processed into little pellets or flakes. These can be put to a variety of new uses – notably the manufacture of synthetic fabrics such as microfleece, now widely used in clothing.

**Green awareness**
*Across the European Union, the amount of plastic in bottles has been reduced by 30 per cent since the 1980s, while 70 per cent less plastic is used in carrier bags.*

# Nitrogen fertiliser 1909

Nitrogen-rich soils are excellent for cultivation, as they are extremely fertile. It was therefore a natural development for people to try to replicate such conditions. By mixing two gases (nitrogen and hydrogen) over an enriched iron or ruthenium catalyst, the German chemist Fritz Haber managed to synthesise nitrogen in the form of ammonia. This substance was then oxidised to produce nitrates and nitrites, used in the manufacture of fertilisers, refrigerants and explosives. In 1909 Haber lodged a patent for his process for extracting nitrogen from air. His colleague Carl Bosch helped him to scale up the process to industrial-level production. Haber was awarded the Nobel prize for chemistry in 1918 for his work in this field.

# Neon lighting 1910

**Art Deco splendour**
*Bright fluorescent signs adorn the frontage of the Hollywood Theater in Los Angeles. Fluorescent tubes use less energy than incandescent bulbs and last 10 to 50 times longer.*

The neon tube was invented in 1910 by a French physicist and chemist, Georges Claude, who built on 19th-century work by earlier scientists, notably Heinrich Geissler, Sir William Ramsay and Morris William Travers. It comprised a simple tube filled with an inert gas at low pressure, through which an electric current was passed. The colour of the light changed depending on which gas was used in the tube: neon produced a red light, argon plus mercury made blue and argon plus sodium yellow. The absence of a filament meant the tube could be any shape one wished. The new light caused a sensation when unveiled at the World's Fair in Brussels and before long bright neon signs were springing up everywhere. In the 1930s neon tubes were replaced by fluorescent, in which the insides of the tubes were coated with a fluorescent substance; again, it was the particular composition that determined the colour. Today, the more ecologically friendly successor is the LED (light-emitting diode) display, which involves no gas, but the term 'neon lights' is still used for illuminated signs.

## UP IN LIGHTS

Britain's first commercial neon signs, advertising Bovril and Schweppes, were turned on in Piccadilly Circus in 1910. In France, the first commercial neon sign was sold to a French barber in 1912 by Georges Claude's business partner Jacques Fonseque. Car salesman Earle C Anthony put up America's first neon sign on his Los Angeles showroom in 1923.

# The sidecar c1910

In 1890 the French engineer Jean Bertoux invented a single-wheeled 'side seat' that could be attached to a pedal cycle. It enabled a passenger to be carried in relative comfort while also serving to stabilise the whole unit.

## SIDECARS IN MILITARY SERVICE

The rugged motorcycle and sidecar, which could negotiate all terrains, saw widespread service in both world wars. Often a light machine-gun was mounted on the sidecar; this column of sidecar machine-gunners (below) was on parade in London during the 1930s.

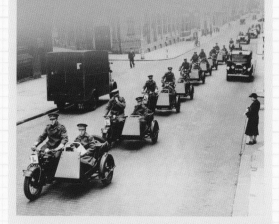

In 1902 an early prototype of the modern motorcycle sidecar was produced in Britain by Mills & Fulford. The following year, the Graham brothers patented the sidecar in both Britain and France. On the Continent, the motorbike manufacturers Johann Puch in Austria and René Gillet in France introduced their own models, but throughout the 1910s Britain dominated sidecar production. Mills & Fulford, Burbur and Watsonian were the leading manufacturers.

Sidecars were extremely popular in Europe in the interwar period and the years of post-war austerity, but suffered a steep decline when affordable small cars came on the market.

**Family outing**
*A typical motorbike and sidecar of the 1930s. Sidecars were mostly made of sheet metal, but there were also wickerwork models and even luxury versions with carpets.*

# Automatic gear box 1910

German engineer Hermann Föttinger (1877–1945) was chief engineer at a shipyard in Stettin on the Baltic. In this capacity he devised a fluid coupling system, which after further development formed the basis of the first automatic gearbox for automobiles. In the days before synchromesh, manual gear changes were a rough business; Föttinger's system made changing gear far smoother and dispensed with the need for a clutch. In Britain in the early 1930s, Leyland Motors adopted fluid coupling, but by 1933 went over to using hydraulic torque converters in many of Lion and Titan buses and in railcars. These devices eliminated the need to change gear every time the bus slowed down and produced a smooth ride. Marketed as 'gearless buses', such vehicles were ideal for use in places without steep hills. Around the same time, the Daimler company developed a semi-automatic transmission system, while in 1939 General Motors invented the Hydra-Matic Drive. After the Second World War automatic gearboxes underwent further improvement, with fluid coupling eclipsed by torque converters. Automatic transmission became the norm for cars in the USA, but European drivers were resistant, remaining loyal to manual stick-shift gearboxes.

## COMPROMISE SOLUTION

Faced with resistance to automatic transmission in Europe, some car manufacturers suggested using hybrid gearboxes, combining automatic and manual functions.

# Unravelling the mystery of chromosomes

The early 20th century saw great strides in our understanding of evolution, thanks to American embryologist Thomas Hunt Morgan and his studies of mutations in fruit flies. Morgan demonstrated that genes are carried on chromosomes and form the basis of heredity. Research into genes would ultimately lead to the discovery of DNA and screening for genetic disorders.

**The human melting pot**

*Children inherit their physical characteristics, such as skin colour, hair type and eye colour, from their parents (top right). But what happens in cases of mixed race? Are certain traits dominant? And if so, over how many generations? These are some of the questions that occupy geneticists.*

The zoologist Thomas Hunt Morgan, Professor of Experimental Zoology at Columbia University in New York, took a special interest in the mechanisms of heredity. He was a fierce critic of Mendelian theories of inheritance, but was greatly drawn to the ideas of the Dutch botanist Hugo de Vries, the man who had largely been responsible for rediscovering the work of Gregor Mendel. From his researches into herbaceous plants, de Vries had concluded that abrupt changes – or 'mutations' – can occur spontaneously from one generation to the next.

Morgan decided to monitor such changes, but unlike de Vries or Mendel, who both conducted their studies with vegetables, he chose to use an animal as his model. Because mice and rats reproduced too slowly, he conceived the novel idea of using fruit flies (*Drosophila melanogaster*), which had ideal characteristics for his purposes.

## White-eyed mutant

In 1910, after breeding several hundred generations of fruit flies, Morgan finally got a mutation. Among his stock of red-eyed flies, suddenly there appeared a male with white eyes. He proceeded to cross this mutant with a red-eyed female: all the first-generation descendants of the union had red eyes. But in the next generation, white-eyed individuals appeared in a ratio of 1:3.

Initially, this finding seemed to confirm the dominant character of red eyes and the recessive nature of white ones, in line with Mendel's laws of heredity established 45 years earlier. But Morgan noticed something strange: the white eyes occurred only among males. He concluded that the determining factor for this physical trait must be the X chromosome. First identified in 1891 by German biologist Hermann Henking, the X chromosome had been found to be present singly in each cell in

---

### THE FRUIT FLY – A MODEL ORGANISM

First popularised for laboratory use by Thomas Hunt Morgan, the fruit fly is still one of the most studied organisms in biology. Easy to raise under lab conditions, the insect has a number of characteristics that made it ideal for Morgan's purposes: its reproductive cycle lasts around 10 days; it is extremely fecund, laying up to 100 eggs a day; and it only has four pairs of chromosomes. Furthermore, despite many differences, fruit flies and humans share almost 60 per cent of their genes. This genetically modified – and hugely magnified – example (right) has two sets of wings.

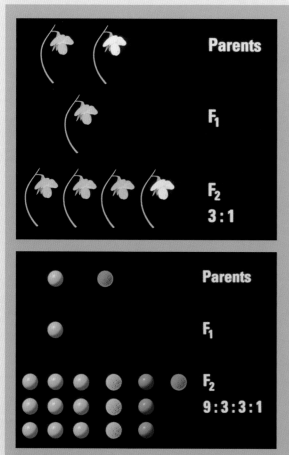

**From parent to offspring**

*According to Mendel's laws, the inheritance of various physical characteristics can be predicted. From parents that differ in one or two traits, the first-generation hybrids (F1) display uniformity of the dominant trait. In the second generation (F2) recessive traits resurface in predictable proportions.*

# THE BIRTH OF GENETICS

In February 1865 an Austrian Augustinian monk called Gregor Mendel (below) presented the results of experiments in hybridising plants to the Natural History Society in the town of Brünn (now Brno in the Czech Republic). No-one at the time paid his report much heed, but they had just heard the first elucidation of the fundamental laws of heredity.

Mendel had spent many years in the garden of his monastery cultivating strains of pea plants. He found they differed in seven characteristics, each of which took two forms, for instance colour (yellow or green) or texture (smooth or wrinkled). After creating pure crosses, he tried interbreeding plants displaying the two different forms of the same characteristic. The resulting first-generation hybrids all turned out to be identical: crossing smooth and wrinkled peas, for example, produced uniformly smooth hybrids. In this case, the characteristic of wrinkling is called 'recessive' and smoothness 'dominant'. From this, Mendel formulated the 'principle of uniformity', his first law of genetics, which states that if plants that differ in just one trait are crossed, all the resulting hybrids will be uniform in the dominant trait. When he went on to cross the first-generation hybrids with one another, he obtained peas that showed the recessive trait in a ratio of 1:3. This gave Mendel his second law on the segregation of 'alleles' (different factors that control each hereditary characteristic), which states that every characteristic is governed by two factors, one of which is passed at random by each parent to its offspring. Finally, by repeating his experiment with descendants that differed in two or three characteristics, Mendel established his third law, that of independent assortment.

Mendel's work lay forgotten for 35 years, but was rediscovered in around 1900 by Hugo de Vries in the Netherlands, Carl Erich Correns in Germany and Erich von Tschermak in Austria.

males, whereas females were found to have two. The Y chromosome, which completes the male pair, was identified by the American geneticist Nettie Stevens in 1905.

In an attempt to corroborate his findings, Morgan then looked for and found among his fruit flies examples of other spontaneous mutations – rudimentary wings, yellow body colour – carried by the X chromosome. This was the first time that hereditary characteristics had been associated with a particular chromosome. In thus determining that some inherited traits were sex-linked, Morgan cast a new light on Mendel's theories. He formulated the hypothesis that each chromosome contains a collection of small entities known as 'genes', using a term coined by the Danish plant physiologist Wilhelm Johannsen in 1909.

**Mendel's third law**

*The law of independent assortment states that characteristics are inherited independently of one another. The principle is shown here by the presence of all-black cows and one brown-and-white cow among the descendants of two parent cows, one of which is black and white and the other plain brown.*

**Complex mix**
*A few traits, such as blood type, are determined by a single gene, but most inherited characteristics are the result of the action of several genes plus the influence of the environment. It is now an accepted fact that a trait that has been lost for several generations can resurface.*

**Genetic pioneer**
*A portrait of Thomas Hunt Morgan who summarised his findings on genetic inheritance in his 1915 work,* The Mechanism of Mendelian Heredity.

## NOBEL LINEAGE

Thomas Hunt Morgan was awarded the Nobel prize for medicine in 1933. He shared the cash prize with his students, many of whom went on to win Nobel prizes themselves.

## The missing link

When Morgan published his findings in the prestigious journal *Science* in 1910, his peers realised that he had found a missing link that explained the role of chromosomes in genetic inheritance. Prior to this, the function of these features within the cell nucleus – first identified in 1888 by the German neuroscientist Heinrich Wilhelm Gottfried von Waldeyer-Hartz – was unclear. Walter Sutton, an American graduate student at Columbia University, had suggested as early as 1902 that chromosomes might carry

the cell's units of inheritance, but his hypothesis was dismissed by the scientific community.

Morgan did not stop there. With the help of three research assistants – Alfred Sturtevant, Hermann Muller and Calvin Bridges – he highlighted other mutations that were or were not associated with sex chromosomes. The team in the so-called 'Fly Room' at Columbia also described mechanisms by which genes are rearranged, which lay at the heart of genetic diversity. Morgan and his students determined how genes are physically positioned along chromosomes. Their work led to the drafting of genetic maps of chromosomes, the first of which was constructed by the 22-year-old Alfred Sturtevant in 1913.

The success of the Columbia unit inspired other scientists to study fruit flies, in the hope of finally deciphering the inner workings of life and of the evolution of species. Genetics became an intensive field of research. Yet years were to pass before these fundamental discoveries could be applied to combating human diseases. Various illnesses, such as haemophilia, thalassemia and muscular dystrophy, were quickly identified as hereditary, but medical genetics, which advises on inherited diseases and the likelihood of transmission, only developed from the 1950s.

Around the same time that Morgan was conducting his studies, the eugenics movement began to gain ground. It had been founded in Britain in

## A FUNDAMENTAL DISAGREEMENT

In the early 20th century, Mendelians like de Vries were convinced that his account of how inheritance works was correct and that species evolved through a series of major leaps, caused by mutations. This principle of 'discontinuous variation' was opposed by Darwinists, who maintained that Mendel's laws did not apply to certain complex traits and that evolution was a gradual, continuous process. The debate continued well into the 1930s, when it was finally resolved by the so-called 'modern synthesis' of evolutionary biology. Stimulated by studies in population genetics, this showed that Mendelian genetics was consistent with natural selection and gradual evolution. In other words, both were right.

the late 19th century by a half-cousin of Darwin, the biometrician Francis Galton. Now its most radical advocates coopted the new science of genetics in support of their theories of racial superiority. Morgan vehemently opposed this unwelcome development.

In the first decades of the 20th century, there still remained much to learn about the function and the nature of genes. In the 1940s George Beadle, a former pupil of Morgan's, demonstrated that genes control the production of proteins, including enzymes, within cells. Oswald T Avery, another Columbia graduate, discovered that DNA is the material of which genes and chromosomes are made. This paved the way for the British biologist Francis Crick and the American mathematician James Watson, building on work by Rosalind Franklin, to reveal the double helix structure of the DNA molecule in 1953 – in their words, 'the secret of life'.

**Under the microscope**
*Chromosomes are long DNA molecules that carry genes. Cells in the human body each carry 22 pairs of X-chromosomes (left), plus one pair of sex chromosomes – XX in women and XY in men.*

**Building blocks of life**
*Changes in the helical structure of DNA, which come about as a result of an exchange with another DNA molecule, lie behind the great diversity of life forms.*

# VITAMINS – 1911
# Essential elements for health

It was the Polish-born biochemist Casimir Funk who first revealed the vital role played by vitamins in human metabolism. His work opened up a new field of medicine aimed at preventing and curing illness through diet.

In 1911 the young biochemist Casimir Funk was working at the Lister Institute of Preventive Medicine in London. He was particularly interested in beriberi, a serious disease of the nervous system and heart that was then ravaging the Far East. In some parts of Asia, half of all infants were dying of this mysterious illness. Some physicians took the view that it was an infectious disease. Funk, though, was intrigued by research conducted in the late 19th century by the Dutch scientist Christiaan Eijkman during his time as a prison doctor on the island of Java in the Netherlands East Indies.

Eijkman had shown that in both chickens and humans beriberi could be combated by a diet of unpolished rice. This differed from polished rice in that the grains retained the hull of rice bran, leading Eijkman to conclude that polished rice contained a toxin, whereas the bran coating contained an antidote. Funk would demonstrate that Eijkman's assumption was wrong: in fact, the substance that cured beriberi was present within the bran itself.

**Nutritious drink**

*The French chocolate drink Banania, introduced in 1912, was rich in vitamin $B_2$ (riboflavin), which promotes growth. It was given to children and also promoted as a pick-me-up for troops at the front in the First World War.*

**Good advice**

*This little rhyme has long been used to encourage children to eat fruit. It is based on sound fact – apples are a good source of vitamins A, C, E and B.*

## The first 'vital amine'

Funk discovered that an amine (an organic compound derived from ammonia) was the element within the rice husk that countered beriberi. He named it 'vital amine', which was soon shortened to 'vitamin'. The particular vitamin in question was $B_1$, or thiamin, which would not be definitively isolated until 1927 and was finally synthesised in 1936. Funk may not have known the specifics, but he had nevertheless opened up new horizons for the understanding of human metabolism. From 1912 on, he developed a theory that many other illnesses such as scurvy and pellagra, long since identified as particularly prevalent among certain groups of people, might be attributable to dietary vitamin deficiencies.

## Building up knowledge

In around 1913 vitamin A was identified by two groups of American researchers working independently. They showed that this substance, which is found in fatty foodstuffs such as milk, butter and cheese, was essential for good nocturnal vision. Their findings vindicated the Greek physician Hippocrates, who some 2,000 years earlier had treated eye problems by prescribing calves' liver, which

---

## WATER-SOLUBLE AND FAT-SOLUBLE VITAMINS

Vitamins are categorised by the materials in which they dissolve. There are two types: water-soluble and fat-soluble. The water-soluble group includes the B-complex (thiamin, riboflavin, vitamin $B_{12}$) and vitamin C. The body only stores these to a limited extent, with any excess being excreted in urine, so they need replenishing on a daily basis through food. The body can store reserves of fat-soluble vitamins such as A,D, E and K.

---

AN APPLE A DAY KEEPS THE DOCTOR AWAY

ALIMENT COMPOSÉ DE CACAO, SUCRE, FARINE DE BANANE ET CRÈME D'ORGE

BANANIA

SURALIMENTATION INTENSIVE
BANANIA
RECONSTITUANT ÉNERGIQUE
CONTENANT
TOUTES LES MATIÈRES INDISPENSABLES
À LA VIE
ADMINISTRATION
48. Rue de la Victoire, PARIS

**Bursting with vitamins**
*Current government health advice recommends eating five pieces of fruit or vegetables a day.*

was also rich in vitamin A. Meanwhile two Norwegians, Axel Holst and Theodor Fröhlich, had discovered that guinea pigs could be cured of scurvy by a diet of fresh vegetables and fruit. Two centuries had passed since Dr James Lind of the Royal Navy had established the curative effect of high doses of citrus fruit juice for sailors suffering from scurvy, and yet the disease was still common among sailors and others with no access to fresh fruit and vegetables. Vitamin C was synthesised in 1932, when the link between its deficiency and scurvy was finally established. Gradually, 13 vitamins were identified, enabling doctors to treat and prevent several serious illnesses through simple dietary regimes.

Vitamins – which, with only a few exceptions, can only be obtained from fresh food – had been shown to be indispensable for the healthy functioning of the human body. Investigations into the properties and effects of vitamins are still ongoing today, as these substances continue to yield new findings. Recent research has shown that vitamin D – whose efficacy in combating childhood rickets is well known – can also help to prevent certain cancers, type 1 diabetes, cardiovascular diseases and seasonal affective disorder (SAD). Scientists have also found that large doses of vitamin D, which is formed in the presence of sunlight, can reduce the risk of bone fractures and osteoporosis in older people.

**A helping hand for health**
*A nurse distributing vitamin supplements to children in a hospital in the south of France in 1941. In Europe during the Depression and the two world wars of the 20th century, vitamin deficiencies in children were usually the result of insufficient food intake. Nowadays, they are often caused by a preference for junk foods, which contain excessive fat and sugar, over fruit and other fresh produce.*

# The heirs of Icarus

The men and women who first seized the opportunity to reach for the skies hailed from a wide variety of backgrounds. Uniting them all was an unshakeable self-confidence, a soaring vision and a stubborn sense of purpose. Many of these early pioneers had multiple skills, designing, building and test-piloting their own flying machines.

**Where it all began**
*Wilbur Wright testing a glider in 1901 (above right). The photograph below shows the Wright brothers in 1910 (Wilbur is on the left). They took turns piloting the Flyer, so it was pure chance that Orville was at the controls for the historic first powered flight on 17 December, 1903.*

At the turn of the 20th century, there was no such thing as a typical aviation pioneer. The men who dared to turn the myth of Icarus into reality were an ill-assorted bunch drawn from across the whole spectrum of inventors. There were methodical theoreticians, solitary tinkerers who dreamed up bizarre contraptions in their sheds, engineers from technical colleges with a streak of the reckless daredevil in them and the idle rich seeking to spice up their lives. Then there were the industrialists speculating that flying would be the next big thing.

## Painstaking and tenacious

Wilbur Wright, who was born in Milville, Indiana, on 16 April, 1867, was one of the first of these innovators to grasp the necessity of mastering flight through a systematic study and theoretical knowledge. He realised that if he was going to go down in history as the person who gave humans wings, then he would need to eliminate the element of chance. After testing a glider with a

wingspan of 6.6 metres in 1901, it struck him that much of the data on which he was basing his work – notably the calculations and tables of the German glider pioneer Otto Lilienthal – simply was not sufficiently accurate. He set about gathering his own data with the aid of a homemade wind tunnel. The heavily modified gliders he constructed in 1902 and 1903 were the fruits of the data he collected from this series of experiments. He made the wings much longer and narrower than before, since this shape proved itself the most efficient in the wind tunnel. Working with his brother Orville, he also steadily improved the controls of their machines, which they equipped with a rudder, a wing-warping system and finally joysticks. In the process they laid the foundations of aviation and, having mastered the art of gliding, the Wright brothers set about fitting their machine with an engine.

## The flying Frenchman

The leading aircraft manufacturers in the world today owe their success to a history of research and development that goes right back to the very beginnings of flight. At the first air show, held at the Grand Palais in Paris in 1909, no fewer than 17 aeroplane-makers displayed their wares. Prominent among them was the engineer Louis Blériot. Born in 1872 in Cambrai in northern France, Blériot graduated from technical college at the age of 23 and embarked on a career making acetylene headlamps for automobiles. He built his first aeroplane in 1905 and was soon dubbed 'The Prince of Bad Luck' by the French press for his frequent crashes. That luck changed in July 1909 when he became the first man to cross the Channel in a heavier-than-air machine.

After this exploit Blériot gave up flying to concentrate on aircraft manufacture. Orders came flooding in. He designed 34 further

### COMMERCIAL FAILURE

Despite earning a million dollars, the Wright brothers failed to fully exploit their early lead in aviation. They dissipated their energies in fruitless lawsuits against other manufacturers and were slow to develop the original Flyer, whose tricky handling made it unsuitable for serial production. After Wilbur died of typhoid in 1912, Orville found it impossible to carry on the business alone and sold his stake in the firm. A much-respected doyen of the aircraft industry, he finally passed away in 1948.

**Historic flight**
*Louis Blériot arriving at Dover in July 1909 (above). Later that year, the Aero Club of France issued him with the world's first pilot's licence.*

prototypes and sold hundreds of his monoplanes prior to the outbreak of war. He bought out the SPAD aircraft company in 1914 and went on to construct several thousand aircraft for the military, which with the impending conflict became the main client for all aeroplane-makers. Two exceptional fighter planes – the SPAD VII and SPAD XIII – accounted for most of the factory's output: more than 12,000 were built, by far the largest production run of any aircraft in the First World War.

At the Armistice, Blériot was the largest aeroplane manufacturer in the world. His interest in aviation remained undimmed and he explored the feasibility of a commercial air service across the Atlantic. Despite the Depression, he continued to develop new planes and seaplanes, many of which were successful. Even so, his company began losing money and in August 1936 his factories were nationalised. Exhausted and suffering from a heart condition, Blériot died that same year.

## A visionary engineer

Early advances in aviation were made possible through the work of brilliant technicians such as the Frenchman Léon Levavasseur. Born near Cherbourg on 7 December, 1863, this son of a naval officer began designing aircraft engines in 1902. The most famous of these was the 'Antoinette', which became the most widely

**Air ace**
*American fighter pilot Eddie Rickenbacker at the controls of his SPAD XIII in 1918 (above).*

**Elegant craft**
*Hubert Latham's Antoinette IV in flight in 1909 (below). Latham attempted to cross the Channel in this machine six days before Blériot achieved his historic feat.*

used aero engine in Europe before 1910. It was water-cooled, with eight cylinders arranged in a 90° 'Vee' and direct fuel injection. Originally generating 24hp, the Antoinette's output was boosted to 50–60 hp in the final version.

## Developments in Britain

Foremost among British pioneers of the period was Thomas Octave Murdoch Sopwith, who in 1910 flew a machine of his own design between Eastchurch in Kent and Tirlemont in Belgium, a distance of 169 miles. Frederick Handley Page established an aeronautical engineering company in 1908 and went on to design many of the British military aircraft used in the First World War. That same year the Short brothers, Eustace and Oswald, founded what was to become the world's first aircraft factory. They created Shellbeach Aerodrome in the Thames Estuary in 1909. The following year they moved to Eastchurch in Essex, along with the Royal Aero Club. The Short-Dunne 5, designed by John W Dunne, was built there – it was the first tailless aircraft

### LEVAVASSEUR'S *MONOBLOC*

In 1911 Léon Levavasseur unveiled the Antoinette Military Monoplane, or *Monobloc*, at the army aircraft trials held in Reims. With its low-slung cantilever wings and boxy fuselage housing a 60hp V-8 engine, the fuel tank and three occupants, it was a genuinely revolutionary design, way ahead of its time in its attempted streamlining. But at 1,250kg fully laden, it was too heavy and underpowered for its weight. It managed only a few short hops at the trials, far short of its intended cruising speed of 150km/h.

Flying the Führer
*Hitler's Junkers Ju52 transport aircraft in flight over Nuremberg.
Leni Riefenstahl's 1934 propaganda film* Triumph of the Will
*featured shots from this plane as it carried the Nazi leader
to the 1934 Nuremberg rally.*

## WOMEN IN AVIATION

**M**any women won renown for their flying exploits in the interwar years. They include Amelia Earhart, the German Hanna Reitsch (the first woman test pilot) and the British pilot Amy Johnson, who in 1930 became the first woman to fly solo from Britain to Australia.

## JUNKER'S *METAL PLANES*

**T**he German aeroplanemaker Hugo Junkers rolled out the first all-metal commercial aircraft in 1919, the F-13 monoplane. It was snapped up by budding airlines and airmail services around the world, including in China, Africa and Australia. The aluminium framework was covered in a duralumin skin. The larger and more advanced Ju52, built in the same way, was introduced in 1930. Upgraded to a trimotor arrangement in 1932, the 'Auntie Ju' was used by the German national airline Lufthansa, and saw extensive service with the Luftwaffe as a troop transport during the Second World War.

to take to the air. In 1911 the Short brothers created the world's first successful twin-engine aircraft, the S.39 or Triple Twin.

### German innovator

Another great name of early aviation was the groundbreaking designer Hugo Junkers, an academic-turned-aeronautical engineer, who was born on 3 February, 1859, at Rheydt in the Rhineland. It was Junkers who introduced the cantilever wing, unsupported by bracing struts or cables. In 1915 he built the J-1, the first all-metal aircraft, nicknamed the 'Tin Donkey'. His designs for metal fuselages were based on sheets of corrugated duralumin, an alloy invented by his fellow countryman Alfred Wilm in 1906. Duralumin was far stronger than pure aluminium, while retaining its lightness. Junkers was responsible for some of the most successful aircraft of the interwar period, notably the Ju52. A socialist and pacifist, he was dispossessed and placed under house arrest by the Nazis in 1934; he died the following year.

Portrait of a pilot
*In 1932 the American pilot Amelia Earhart (above) became the first woman to fly solo across the Atlantic. Five years later, she disappeared without trace over the Pacific.*

# The world takes wing

The first intrepid aviators who lifted off in heavier-than-air machines were all too aware of the huge gulf between the urge to fly and the strenuous, dangerous business of keeping a plane aloft. But thanks to the risks they took in testing out new technology, aircraft quickly improved until they were safe and reliable enough to carry fare-paying passengers.

Instrument panels on the earliest aeroplanes were rudimentary in the extreme. To get their bearings and judge conditions as they flew, pilots simply kept their eyes peeled and looked around them. On clear days, using observation with the naked eye, it was relatively easy to tell if the plane was flying straight and level, and to judge manoeuvres involving climbing, descent and banking with reasonable accuracy. If pilots lost their way, all they needed to do was drop down low enough to pick up a railway line, then follow it to a station and read the large signs on the platforms. But the moment a plane flew into cloud, visibility disappeared and the risk of an accident increased dramatically.

## Breakthrough in aerial navigation

When making his cross-Channel flight in 1909, the only instrument on board Blériot's Type XI monoplane was a fuel gauge. But before long, cockpits began to fill with instruments that made flying a less haphazard affair.

From 1914 onwards, most aircraft were equipped with an airspeed indicator, an altimeter, a rev counter, a thermometer, a clock and a roller map. Radios were introduced very early on, but their usefulness was limited as aerials that would allow anything but very short-range communication were too heavy. The answer was a radio receiver and directional antenna linked to a dial to tell the pilot or navigator where incoming radio signals were coming from. These radio compasses were fitted with a small electric motor that kept the antenna turning constantly.

As night flying became more commonplace, well-used air routes were marked with beacons along their entire course. But again, this system only really worked in fair weather conditions.

**Wings over London**
*The first airport serving London was at Croydon, which opened to commercial traffic in 1920. In 1933, to cater for the growth in air travel, the Air Ministry approved commercial flights from Gatwick. Heathrow opened in 1946.*

## 'THE MAIL MUST GET THROUGH!'

The success of the aeroplane led quickly to the introduction of airmail. America was in the vanguard, and on 17 February, 1911, Fred Wiseman made the first official airmail flight, carrying three letters between Petaluma and Santa Rosa, California. Next day, the French pilot Henri Pequet carried 6,500 letters a distance of 13km from Allahbad to Naini In India. That same year, on 9 September, the first scheduled airmail postal service was inaugurated between Hendon, north London, and Windsor.

British troops stationed in Germany after the First World War received mail by air. In the 1920s the RAF was responsible for developing routes for airmail to the Middle East. The French airmail service Aéropostale was inaugurated in 1918 at the instigation of Pierre Latécoère, a military aircraft manufacturer who had ambitions to expand into commercial aviation. In September 1919 he launched a regular service, called simply 'La Ligne', ('the Line'), between Toulouse and Morocco, which he gradually extended to Dakar in Senegal (then known as French West Africa). Among its early pilots was Antoine de Saint-Exupéry, who later became famous for books such as *Night Flight* and *Southern Mail*, recounting his flying exploits, and his children's novella *The Little Prince*. Conditions for long-distance mail pilots were gruelling – they had to contend with engine failures, appalling weather and even attacks by hostile tribesmen. Some downed airmen were captured and held to ransom. The heroes of 'the Line' endured such hardships guided by the service motto: 'The mail must get through!'

### Flying far and wide
*A Potez-25 biplane of the Aéropostale service flying over the Andes in 1929 (top). The company promoted its services with posters like this one published in Brazil (above).*

### Famous pilots
*Antoine de Saint-Exupéry (on the left) and Henri Guillaumet pose with an Aéropostale Latécoère-28 monoplane, mainstay of the French airmail service.*

With the coming of radio navigation, bonfire beacons were replaced with the radio variety. On 24 September, 1929, US Army pilot Lt James Doolittle achieved the first totally 'blind' flight in a two-seater plane with the cockpit shrouded in tarpaulin. This was made possible by highly accurate navigational equipment that was soon fitted to all aircraft.

## The growth of commercial aviation

Immediately after the First World War, many still doubted whether aeroplanes would ever be capable of transporting cargo or passengers on a regular basis. Yet within three months of the Armistice, commercial flights were being made. On 5 February, 1919, Deutsche Aero Lloyd (precursor of Deutsche Luft Hansa) made the world's first commercial flight, between Berlin and Weimar. Three days later Lucien Bossoutrot, a pilot for the French Farman company, flew a converted bomber carrying 11 paying passengers from Paris to London. The first British company to operate regular international flights was Aircraft Transport

**Precious cargo**
*Unloading gold bullion from an Imperial Airways plane at Croydon Airport in 1926. Imperial Airways was founded in 1924 to serve all outposts of the British Empire.*

### STRAIGHT AND LEVEL

The gyroscope, which lies at the heart of the gyrocompass, was first studied by the French physicist Léon Foucault in 1852. In 1908 Hermann Anschütz-Kaempfe developed a gyroscopic compass for use by the Imperial German Navy. Shortly afterwards, in around 1912, the American inventor Elmer Sperry patented a similar stabilising device for aircraft. One gyroscope registered the roll of the aircraft and another the pitch. These then acted on the controls via a pneumatic system, keeping the plane straight and level. Automatic pilots were introduced in 1947.

**More than a wing and a prayer**
*By 1933, when this edition of the American monthly magazine* Fortune *appeared, pilots had an array of sophisticated instruments in the cockpit to help them to navigate and fly.*

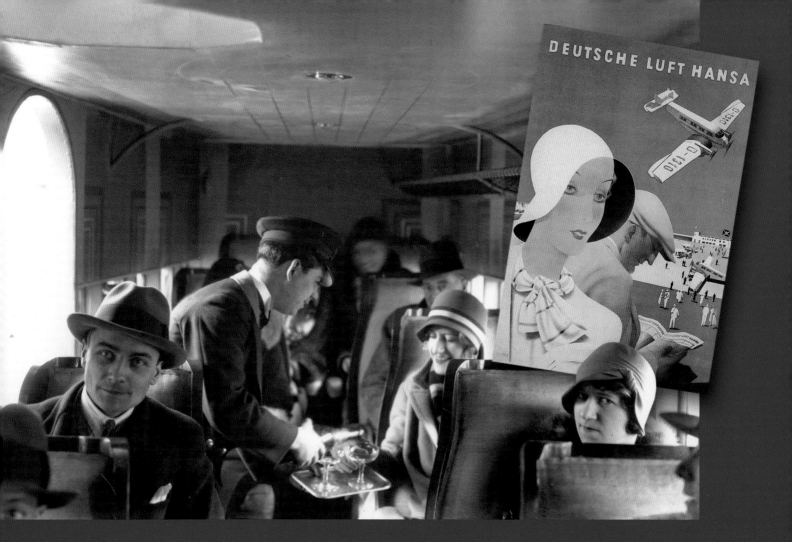

## TRANSATLANTIC FLOP

**German engineers developed the huge *Do-X*, a 12-engined, 55-tonne Dornier flying boat, for transatlantic travel. But in 1930–1 its proving flight from Friedrichshafen to New York was plagued with mishaps. The leviathan was underpowered and its poor performance doomed it to failure.**

and Travel Limited, founded in 1916 by George Thomas using military aircraft. On 15 July, 1919, it began flights between RAF Hendon and Paris Le Bourget, charging £21 per passenger for the journey of two and a half hours. Regular flights began the next month, making it the world's first daily plane service.

Yet for now, there were natural limits to aviation growth. The only people in a position to take advantage of the new mode of transport and the freedom it offered were the wealthy. Even then, a journey by plane was an endurance test. Cabins were cramped, cold and draughty as the slow, low-flying planes were constantly buffeted by the wind. And there were still many practical problems involved in flying that required technological solutions. The Germans invented radio-guided landing and the French the radio compass.

Airlines began to proliferate around the world, but they were only viable with state subsidies. From the early 1930s the US government began actively supporting the development of new transport aircraft. In time, this enabled manufacturers like Douglas and Lockheed to corner the global airliner market. Meanwhile, in 1938, the Boeing company of Seattle built the first airliner with a pressurised cabin. The first transatlantic flight to carry passengers was made on 28 June, 1939, from Port Washington to Marseilles, by a four-engined Boeing 314 'Clipper' flying boat. British *Empire*-class flying boats soon followed suit. But with the intervention of the Second World War, commercial air transport was put on hold, only taking off again in the late 1940s. Expansion into a mass market came 20 years later.

**Luxury service**
*Passsengers on an Air-Union flight from Paris to London in 1929 were served champagne to mark the 10th anniversary of the route. The Air-Union group was formed in 1923 from the merger of smaller companies such as Bréguet and Blériot. In 1933 it became Air France. The poster for Luft Hansa dates from the 1920s.*

---

### THE CONCORDE OF ITS DAY

In 1931 the American manufacturer Lockheed introduced its *Orion* monoplane, the first commercial transport aircraft with a retractable undercarriage. The following year Swissair bought two *Orions*, which it painted a vivid scarlet, earning them the nickname 'Red Dogs'. The *Orion* carried just four passengers in cramped conditions, but cruising at almost 300km/h – twice the speed of any other plane at the time – it still helped to revolutionise the European airline industry.

# CONTINENTAL DRIFT – 1912

# From one to many

**W**hen German meteorologist Alfred Wegener presented his radical theory of continental drift, he was greeted with scepticism by his peers. He spent his life trying to convince the scientific community that the continents once formed a single landmass. Wegener was finally vindicated by the universally accepted science of plate tectonics.

Pangaea – 200 million years ago

How were Earth's continents formed? People in the ancient world believed that the oceans had always occupied the same positions and that landmasses simply rose up out of them. The 17th century, which saw increasingly accurate mapping of Europe and the Americas, witnessed the rise of the so-called 'catastrophist' theories, which claimed that the location and shape of the present continents were the result of sudden, cataclysmic past events. But in a paper delivered to the Frankfurt Geological Society in 1912 – and expanded three years later into a seminal work entitled *The Origin of Continents and Oceans* – the German explorer and meteorologist Alfred Wegener (1880–1930) proposed a revolutionary new hypothesis.

According to Wegener's theory, some 200 million years ago the continents were joined in a single landmass that he called 'Pangaea', and this was surrounded by one vast ocean, Panthalassa. The present configuration of the continents came about as a result of the splitting of this immense original continent into smaller blocks that gradually drifted apart over millennia.

## Maps provide the proof

Other scientists had already raised the idea of continental drift, notably the American geologist Frank Bursley Taylor in 1908, but Wegener was the first to provide supporting evidence for the theory drawn from careful examination of maps. After studying the outlines of Africa and Europe on the one hand and the Americas on the other, he noticed in particular that the eastern coastline of South America looked as thought it would fit with the west coast of Africa, like the interlocking pieces of a jigsaw puzzle. His first inclination was to reject the

**Man of many parts**
*Wegener was a polymath pursuing interests in physics, astronomy and meteorology. In 1930, aged 50, he disappeared while on an expedition to Greenland to monitor Arctic weather. His body was recovered the following year.*

idea, reasoning that it was highly improbable that landmasses could shift across so great a distance. Yet only a year later he returned to the problem, after reading the reports of palaeontologists who had explored both Africa and South America.

These researchers pointed out the striking resemblance between fossils from either side of the Atlantic. To explain this phenomenon, palaeontologists and geologists alike usually resorted to the concept of land bridges, which were thought to have periodically arisen and disappeared as sea levels fell and rose, allowing animals to migrate from one continent to another. But Wegener's theory of continental drift, which by now he was utterly convinced of, dispelled this somewhat fanciful idea. His hypothesis that the continents had once been a single landmass before splitting and drifting apart offered a far more plausible explanation not only for the similarity in the coastal outlines but also for the presence of similar species on either side of a very wide ocean.

## Further confirmation

Wegener then embarked on the long process of assembling more pieces of evidence to support his ideas. He devoted the 1910s and 1920s to studying all the latest findings in geology and palaeoclimatology. He learned, for instance, that there was widespread evidence of glacial sediments from the Permo-Carboniferous

**Continental drift in action**

*Three stages in the formation of Earth's continents (left and below). The landmasses continue to drift at the rate of a few centimetres a year.*

period throughout the tropics. These took the form of rock debris called 'tillites', which were deposited by glaciers and had now been found in South Africa, Madagascar, Arabia, India, Australia and elsewhere. Yet how could these tropical regions – even in the distant past – have been subject to glacial temperatures? For Wegener this was an important new confirmation of his ideas. According to his concept of Pangaea, the regions in question would at one time have been closer to the South Pole and therefore experienced glacial activity.

As the continents separated and drifted north towards the tropics, the glaciers duly melted.

The similarity and continuity of geological formations between the continents supplied him with further proof. He noticed, for example, that rock types and the directions of rock folding were identical across the Appalachian mountain range on the eastern seaboard of North America, the Mauritanides of northeast Africa and the Caledonian range, which spanned the British Isles and Scandinavia. This pointed to a common geological origin.

**In the Cretaceous – 146 million years ago**

**Earth's continents today**

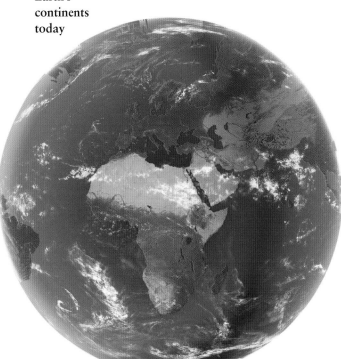

## ONE FOSSIL CLINCHES IT

Among the fossils that Wegener used to support his theory, *Mesosaurus* takes pride of place. This freshwater aquatic reptile evolved at the start of the Permian period, some 280 million years ago. It was around one metre long and had teeth like a modern crocodile or alligator. The first example to come to light – named *Mesosaurus tenuidens* – was found in Griqualand (modern Cape Province in South Africa) by the French palaeontologist Paul Gervais in 1864. Then, in 1908, an almost identical fossil, *Mesosaurus brasiliensis*, turned up in Brazil. Because the animal's physiology meant that it could not possibly have swum across a large body of salt water like the Atlantic Ocean, the only plausible explanation for Wegener was that the continents must once have been joined together.

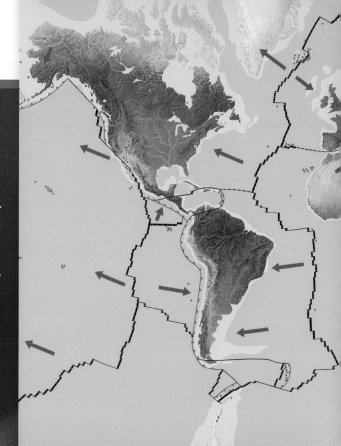

## PLATE TECTONICS

**D**uring the Second World War, Harry Hammond Hess, a geologist by training, became captain of a US Navy transport ship fitted with sonar. As his ship plied the Pacific, he collected huge amounts of data on ocean-floor profiles. His surveys revealed the presence of massive submarine volcanoes (guyots) and seismic fault lines in the Earth's oceanic crust. From this Hess developed his theory of seafloor spreading, which postulated that the Earth's crust is gradually moving away laterally from long, volcanically active underwater ridges.

In the late 1960s, the findings of Hess, Wegener and other researchers were combined into the broader theory of plate tectonics. These plates – 15 were identified in total – consist of a lithospheric mantle overlain by two kinds of crustal material: oceanic crust and continental crust. The plates themselves ride on an underlying layer of magma (molten rock) called the asthenosphere some 100–170km (60–100 miles) beneath the Earth's surface.

The movement of the tectonic plates in different directions, by increments of several centimetres a year, occurs either through subduction (where one plate moves beneath the next) or collision. It is this movement between plates that produces volcanic activity and mountain chains. The Himalayas, for example, were formed by the Indo-Australian plate crashing into the Eurasian.

Geologically active
*The archipelago of Vanuatu in the Pacific comprises 80 separate islands. Situated at the boundary of the Pacific and Indo-Australian tectonic plates, it is regularly hit by earthquakes and tsunamis. This, plus three active volcanoes including Yazur (below), makes it an important location for earth scientists.*

Tectonic plates and fault lines
*The fault lines along boundaries between tectonic plates are geologically active zones: most volcanoes (indicated on the map by red dots) lie close to these boundaries. The red arrows on the map indicate the direction in which the plates are moving. The collision zones, called constructive plate boundaries, are where mountain ridges are formed (shown as pale pink lines), whereas the subduction zones, or destructive plate boundaries (violet lines), are the site of earthquakes and volcanoes.*

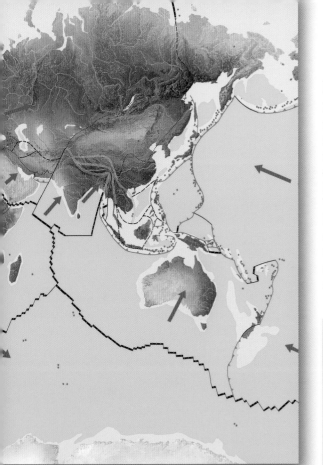

## Belated recognition

It was only after the Second World War, with the beginning of undersea exploration, that the movements of the Earth's crust became a live topic once more. Various hypotheses finally coalesced into the theory of plate tectonics. Formulated in 1968, this owed a great deal to Wegener's ideas. But vindication sadly came too late for Wegener himself. In 1930, while on a scientific expedition to study the ice and climate of Greenland, he was killed as he attempted to supply provisions to two colleagues who were marooned far from base camp.

## A hostile reception

The evidence put forward by Alfred Wegener still failed to convince geologists, who only began to take notice of his ideas in 1922. Their reaction was universally dismissive. There was no questioning Wegener's skill as a meteorologist and balloonist. In 1906, during one of his studies of the Earth's atmosphere, he had smashed the world ballooning endurance record by clocking up a non-stop flight of 52 hours over Germany. Nor was his physical endurance in doubt; he led several arduous expeditions to Greenland to study the polar climate there. But the fact remained that he wasn't a 'real' earth scientist. His critics, like the British geophysicist Harold Jeffreys, even went so far as to cast aspersions on his academic credentials.

Wegener's theory was also hampered by key questions that remained unanswered. What kind of forces were capable of moving masses as heavy as continents? And even supposing they could move, what were they floating on? Wegener put forward various theories, such as the existence of a centrifugal force pushing out from the Earth's core to its surface, or friction by the tides pulling the continents inexorably westwards, but he never really found a satisfactory answer to these questions.

### WHAT CAUSES EARTHQUAKES?

As a result of the systematic installation of seismographs around the Earth in the late 1960s and the formulation of the theory of plate tectonics at around the same time, we know that the great majority of earth tremors occur at faultlines where one tectonic plate abuts another. Earthquakes are caused by the sudden release of energy stored up along the faultline as a result of friction between the plates in the period leading up to the tremor. Around 100,000 quakes are recorded every year, most of which cause no damage and are so slight that people often do not even notice them. But when a large amount of energy is released, the resulting earthquake can be devastating. If this happens at an underwater faultline, as in the tremor that hit Sumatra on 26 December, 2004, it can unleash a tsunami (tidal wave) with catastrophic consequences.

**Major disaster**
*At the San Andreas fault, which runs down the west coast of North America, the Pacific plate is moving against the North American plate at a rate of about 6cm a year. This causes regular tremors in the region, but on 18 April, 1906, it produced a quake that measured 8.2 on the Richter scale and devastated San Francisco (left).*

# Maglev technology 1912

In 1904 the American scientist Robert Goddard, who was to make his name as a rocket scientist, wrote a paper while still a freshman at Worcester Polytechnic Institute in which he proposed a frictionless form of travel by using electromagnetic repulsion to lift train carriages off their rails. The trains would travel at fantastic speeds inside a steel vacuum tube. But it was a Frenchman who was destined to develop the theory further. On 15 March, 1912, the *Daily Argus*, a newspaper serving the New York suburb of Mount Vernon, carried a report on a new technological marvel which, it confidently claimed, 'would allow mail to be sent from here to Boston in less than an hour'. The subject of the article was the latest invention of the French-American electrical expert Émile Bachelet – a cigar-shaped metal tube one metre long, which he had propelled down a track at the astonishing speed of 500km/h (310mph). As in Goddard's idea, it relied on the simple principle of like magnetic poles repelling one another. Bachelet installed powerful electromagnets in both the tube and along the rail. Some of them were used to levitate the vehicle off the track, while others drove it forward. His patent outlined how the magnets would only become charged with electricity as the vehicle passed along the rail. Bachelet originally conceived just a small device intended to carry letters, but quickly realised that he could upscale it into a full-sized train.

With British finance, Bachelet established a laboratory in London in 1914. The demonstrations he staged there earned him instant fame, but his plans never came to fruition. It was not possible at the time to generate a current strong enough to power a train. It would also have required a whole new track infrastructure next to existing railway lines.

After the First World War, the German engineer Hermann Kemper revived the idea and filed a patent for a magnetic levitation train in 1934. Yet it was the 1960s before the Japanese built the first prototype of a Maglev train, which broke a series of speed records, the last being 581km/h (361mph) in 2003. Research is still in progress. The Germans began working on their rival Transrapid system in the 1970s, but the project was abandoned in 2008. The only commercial Maglev service currently in operation is the German-built Transrapid between the Chinese city of Shanghai and Pudong International Airport.

**Frictionless travel**
*The German Transrapid 08 at the Emsland test facility in 2004. The high cost of constructing Maglev technology has so far hampered development of this potential mode of transport.*

## THE WORLD'S OLDEST MONORAIL

In 1901 a revolutionary monorail system was built by the engineer Eugen Langen in the city of Wuppertal in northern Germany – and is still running to this day. Technologically less advanced than a magnetic levitation railway, the passenger carriages of the *Schwebebahn* ('suspended railway') hang below the track, which is supported by steel gantries. Electric motors with an output of 600W DC are mounted directly on the driving wheels. Trains on the Wuppertal monorail travel along the 13.3km track at speeds of up to 50km/h (30mph). In more than a century of operation, it has suffered only one serious accident. which happened in 1999.

# Stainless steel 1913

Harry Brearley, a metallurgical researcher working in laboratories run jointly by Sheffield steelmakers John Brown & Co and Thomas Firth & Sons, developed an important new product in August 1913. Brearley had been commissioned by an arms manufacturer to devise a corrosion-resistant alloy that would overcome the problem of excessive wear on the internal surfaces of gun barrels. Using a mixture of iron, carbon (0.24 per cent) and chromium (12.8 per cent), he noticed that, unlike conventional steel, his alloy could not be tarnished by nitric acid and was resistant to corrosion. He immediately christened it 'rustless steel' and took out a patent on it.

## An old formula

In fact, stainless steel and chrome alloys that were resistant to acids had been produced in the 1820s both by the English chemist Michael Faraday and by the French metallurgist Pierre Berthier. Many other scientists in Britain, France and Germany attempted to exploit the idea in the following decades, but were thwarted by the fact that their alloys contained too much carbon and as a result were too brittle. Just a few months before Brearley's breakthrough, researchers at Krupp, the giant German armaments manufacturer in Essen, filed patents for a stainless steel containing 18 per cent chrome and 8 per cent nickel, which they called 'Nirosta'.

## Ideal for cutlery

If Brearley was not the first to invent stainless steel, he was the first to produce it on an industrial scale. He quickly spotted the potential of his rustless steel. A metal that resisted nitric acid would clearly shrug off vinegar and lemon juice. Sheffield had long been a centre of cutlery manufacture, and Brearley realised that his product would be ideal for making knives, forks and spoons, not to mention pots and pans. The name 'stainless steel' was suggested by Ernest Stuart, managing director of local cutlery manufacturer R F Mosley. During the First World War, Sheffield's output was given over to the war effort but research on stainless steel resumed in 1918. Nowadays, there are hundreds of different types of stainless steel, designed for many different uses, including precision instruments, car bodywork, furniture and the frames of high-rise buildings.

**Fine dining**
*An early stainless steel tea knife, made in 1915 by Sheffield's most prestigious cutlery maker, George Butler.*

---

### THE IRON PILLAR OF DELHI

In the Qutb complex of ancient monuments at Mehrauli in Delhi, there is a pillar of wrought iron, standing 7 metres tall, which dates from the reign of Chandragupta Vikramaditya (375–413). It has resisted corrosion in the open air for more than 1,600 years. The secret of its longevity without rusting lies in the high proportion of phosphorus in the metal, which has promoted the formation of a thin protective film of iron hydrogen phosphate hydrate on its surface. The pillar is not a stainless steel in the modern sense of the term, as it contains no chromium.

---

**Flashy motor**
*The American automobile industry was quick to make use of the new rust-resistant metal. The best-selling 1938 Buick (right) had an imposing radiator grille and front bumper made of highly polished chrome-plated stainless steel.*

# Crossword puzzles 1913

The word square, or 'magic square', had been known since ancient times. It consisted of a series of words, all with the same number of letters, which when arranged in a square grid pattern read the same across as they did down. A Latin puzzle of this type was found in the ruins of Herculaneum, buried by the volcanic eruption of Vesuvius in AD 79.

## The modern crossword

The rise of the modern crossword is due to an Englishman, Arthur Wynne (1862–1945), who was an editor and puzzle compiler from Liverpool. Wynne created 'magic square' games for syndication to British newspapers, but they never really took off. After emigrating to America in the hope of finding a more receptive audience, he submitted one of his 'word-cross puzzles', as he called them, to the Christmas edition of the *New York World* in December 1913. His puzzle duly appeared as a diamond-shaped grid, with no black spaces. It was an instant hit and within a decade most major newspapers in the USA were publishing a daily crossword. Thousands of dollars were sometimes offered for correct solutions.

**Crossword clothing**
*A seaside postcard from the 1920s shows a promenading couple dressed in outfits inspired by the new puzzles. Even the dog is black and white.*

**All the rage**
*Reflecting the latest craze, this crossword-themed tea dance took place in East Ham in London in 1919.*

**THE D-DAY CROSSWORDS**

In the run-up to the Allied invasion of Nazi-occupied Europe in 1944, various solutions to clues in *Daily Telegraph* crosswords gave British military intelligence serious cause for alarm. From March to 1 June, all the names of the Normandy landing beaches ('Juno', 'Gold', 'Sword', 'Utah', 'Omaha') appeared, alongside top-secret codewords for other aspects of the operation. The compiler, a schoolmaster named Leonard Dawe who had been writing crosswords for the newspaper since 1925, was interviewed by MI5 on suspicion of espionage, but never charged. He claimed he had simply overheard the words from US and Canadian troops billeted near the school where he taught.

## Global success

Back in Britain, meanwhile, crosswords were revived by the journalist Morley Adams. He failed to credit Wynne, but he did strike a chord with the public. Before long crosswords spread to Europe, appearing in both French and German newspapers for the first time in 1925. Cryptic crosswords, whose clues involve wordplay, became especially popular in Britain in papers such as *The Times* and *Daily Telegraph*. Today, millions of solvers worldwide sharpen their wits with crosswords. Some neurologists even claim that this form of brain stimulation can help to stave off the onset of memory-loss disorders and dementia.

# Heparin 1916

In 1916 a young American medical student at Johns Hopkins University in Baltimore discovered a molecule in canine liver cells that could prevent blood from coagulating. The student was Jay MacLean and he named his discovery heparin (from the Greek *hèpar*, meaning 'liver'). The substance was also produced naturally in other animals, such as pigs and humans. Two of MacLean's colleagues, William Howell and Luther Emmett Holt, described the chemical properties of heparin in 1918, but it was 1936 before it was administered for the first time to a patient as an anticoagulant drug. The heparin was extracted from intestinal tissue by a process of proteolysis (digestion by cellular enzymes) before being purified by precipitation. In the 1970s various scientific papers revealed how the drug worked: it was found, for instance, that it inhibited the formation of thrombin, an enzyme that appears in the blood immediately prior to clotting.

## Prevention and cure

The discovery of heparin is widely regarded as one of the major medical breakthroughs of the 20th century. Currently, some 500 million doses of this drug are administered worldwide each year. It is used preventively against thrombosis (the formation of blood clots in the veins and arteries) in permanently bed-ridden patients or for those who are about to undergo operations, and in cancer patients, who are particularly prone to this disease.

Heparin is used in heavy doses to combat deep-vein thrombosis (DVT), pulmonary embolisms and myocardial infarctions (heart attacks). Depending on the specific illness, it is injected either intravenously or subcutaneously. Treatment involves close monitoring of the patient, since an overdose of heparin can cause severe haemorrhaging. However, its effects can be reversed by giving an antidote, protamine sulphate.

**Under the microscope**
*A scanning electron microscope image of blood cells (above) shows healthy, disc-shaped red blood cells with T-lymphocytes (in blue), platelets (orange) and fibrins, proteins involved in blood clotting (green). Also visible is an echinocyte (pink), an altered red blood cell with knob-like projections that is associated with kidney disease.*

---

### A WISE PRECAUTION

Because the intestines of cattle were the first organs to be affected by prions – the infectious agents responsible for BSA, bovine spongiform encephalopathy commonly known as 'mad cow disease' – heparin from cattle is banned in Europe and the United States, where heparin from pig intestines or the synthetic variant are used instead.

---

# Breaking out of the trenches

It was Leonardo da Vinci who first dreamt up the idea of the tank, but it was the British who first made it a reality on the battlefield of the Somme in the First World War. Incorporating some of the most advanced technologies of the day, these lumbering vehicles were soon an indispensable weapon of war.

**Ahead of its time**
*Leonardo da Vinci's notebooks contain this sketch of a circular, tank-like armoured wagon (right), designed at the instigation of the Duke of Milan.*

**Tank offensive**
*A British Mark IV tank in France in 1918. Following the successful deployment of tanks at Cambrai in 1917, the British came to rely ever more heavily on tanks to support infantry.*

By September 1916 the Battle of the Somme on the Western Front had been raging for more than two months and the British had all but lost. But they had one last card up their sleeves: the new Mark I tank. On 15 September, as clandestinely as possible, 32 of these were deployed on a 5-mile section of the front along an old Roman road from the town of Albert to Bapaume. The British hoped their new wonder weapon would manage to punch a hole in the German lines. In the small hours the 28-ton behemoths, armed either with four machine-guns (the 'female' variant) or two 6-pounder (57mm) cannon (the 'male'), crept forward into no-man's land between the two armies. As the bullets from their Mauser rifles ricocheted off the vehicles' armoured flanks, at first the Kaiser's troops panicked and took to their heels.

Inside each tank was an eight-man crew in cramped and appalling conditions. The cabin was dark, while the lack of any suspension over the shell-pocked terrain meant the men were flung about against the thick steel plating. The engine made a fearsome din – it was so loud the only way to attract attention was to bang on the tank's side with a metal object – and churned out so much heat that at times the temperature inside exceeded 50°C. The air was thick with petrol and exhaust fumes, poisonous carbon dioxide and cordite from the guns.

Only nine tanks made it through the German defences on the Somme that day. The rest of the tanks in the British armada ground to a halt before they could reach the enemy trenches, stopped by mechanical breakdown or shellfire. Their armour was only 12mm thick at best and their top speed of 4mph (6km/h) was little more than walking pace. Out of reach of British artillery support, they were gradually picked off one by one and all the captured territory was regained.

## The quest for land-based ironclads

As the opposing armies had dug in on the Western Front in late 1914, the stalemate that ensued, with troops facing each other from lines of heavily defended trenches, forced the combatant nations to try to think of new ways to break the deadlock. Infantry charges were suicidal, as the men were mown down in their thousands by machine-guns. But barbed-wire and countless shell-holes made the terrain impassable for conventional wheeled vehicles. An armoured all-terrain vehicle appeared to hold out a possible solution.

To turn the idea into reality, the engineers first turned their attention to caterpillar tracks, which dated back as far as 1770. By the outbreak of the First World War, caterpillar tracks were already a familiar sight on the

development were agricultural engineer William Tritton, his opposite number in the Admiralty Lieutenant Walter Wilson, and Lieutenant-Colonel Ernest Swinton. Existing tractor caterpillars proved to be unsuitable, so they were widened. Even so, the first prototype tank, nicknamed 'Little Willie', was judged a failure and it was its successor, 'Big Willie' (or 'Mother'), that

'crawler'-type farm tractors made by the British company Hornsby or the US firm Holt; indeed, the Allies used such vehicles to drag their heavy artillery pieces up to the front across rough ground. Under the auspices of the Admiralty, the British set up a so-called Landships Committee to investigate the possibility of making an ironclad that could operate on dry land. From the inception of the project in August 1915, research and construction of the first practicable tank took just a year. Those primarily responsible for its

**Cramped quarters**
*The American skipper and gunner of a 'Whippet' light tank, a British design, in action near Verdun in 1918 (above). 'Splatter masks' like this one (left), made of leather and chain mail, were issued to tank crews to help protect their faces from shrapnel.*

### KEEPING IT HUSH-HUSH

To ensure that the development of an armoured fighting vehicle remained top-secret, the British Admiralty told workers at the factory where the new weapons were being assembled that they were making 'mobile water tanks' for use in desert warfare in Mesopotamia. The name stuck.

## TRIAL AND ERROR

The early days of the tank did not go well. The aim of British engineers was a machine with a top speed of 4mph and a working radius of 20 miles, which could climb a 5ft parapet and cross an 8ft gap. As the first model emerged on 8 September, 1915, its track fell off, a failure repeated 11 days later in the presence of government officials. As a result, William Tritton and Walter Wilson designed a more reliable, wider track, after which development proceeded apace. The first French vehicle, the Schneider CA1, was poorly designed and ineffectual, as was the St Chamond, which was heavy and prone to getting stuck. The French success story was the FT-17, designed and built by the automobile firm Renault.

**American offensive**
*A squadron of FT-17 tanks at the Battle of St Mihiel on the Meuse in September 1918, commanded by Colonel George S Patton of the American Expeditionary Force.*

**Formidable fighting machine**
*The Israeli 'Merkava 3' main battle tank, seen here (below right) on the Golan Heights in 1998, is well-armed and fast, capable of 30mph (50km/h) over even the roughest terrain. To protect its crew, the engine is positioned at the front of the hull.*

established the standard configuration of the first British tanks. These had a low-slung hull, which housed the crew and the engine, set between two huge rhomboid-shaped sections, around which ran the caterpillar tracks. On the flanks of the vehicle were two sponsons, each containing a naval 6-pounder gun (or a machine gun). In the event, this first British tank design remained a one-off; all other such vehicles built in the First World War took their cue from the French Renault FT-17 light two-man tank, which was the first to incorporate a top-mounted turret with full rotation.

After their shaky start, which gave the Germans time to develop anti-tank weapons, tanks came into their own at the Battle of Cambrai. On 20 November, 1917, a force of 324 British Mark IV tanks smashed through the Hindenburg Line, penetrating 4 miles (6km) into enemy territory. In 1918 the victorious Allied offensive was supported by large concentrations of Mark Vs and FT-17s. The success seemed to herald a bright future for the tank; no other weapons system combined such a measure of protection, speed and firepower.

### Lighter and faster

In the interwar period, the trend was for tanks to become lighter and faster. One radical new design was the French Somua S-35 of 1934,

**Teutonic tank**
*The ungainly A7V (above), nicknamed the 'Monster' by its crews, was the only German tank type produced during the First World War.*

which weighed between 13 and 20 tons and could travel at 18mph (30km/h). Welding of hulls rather than riveting made tanks stronger and safer (rivets could fragment under fire and fly around the interior). Sloped armour also made an appearance, although it only became widespread in the Second World War.

During that first deployment of tanks on the Somme, a Mark I that got separated from the infantry advance managed to overrun a German trench by coordinating its attack with a low-flying British aircraft. This chance tactic was not repeated at the time, but in the 1940s the Germans made it the cornerstone of their *Blitzkrieg* offensive against the Low Countries and France, with massed armoured divisions, motorised infantry units and ground-attack aircraft like the Ju87 'Stuka' dive-bomber working in close concert. Up to 1942 this tactic, which was made possible through the installation of radios in tanks, brought the German army some spectacular successes.

Tank design came on in leaps and bounds during the Second World War. The thickness of armour-plating, for example, increased from 15mm in the *Panzerkampfwagen III* (the standard German army tank in 1940) to 40–80mm in the *Panther*, reaching 80–110mm by the end of 1944 in the *Tiger II*. The many roles that tanks were called upon to perform saw the development of light, medium and main battle tanks, as well as highly specialised variants such as amphibious, mine-clearing, flamethrower or bridging tanks.

Cold-war tanks such as the Soviet T-62 were even faster and better armed, with 105mm guns or larger becoming the norm, compared to 75mm cannon in the Second World War. They also became more genuinely multi-purpose and standardised. But the great advances in anti-tank weapons made in the 1960s – such as high-explosive anti-tank rounds, armour-piercing shells and wire-guided missiles – meant that tank manufacturers were constantly trying to stay one step ahead. Answers were often found in new technologies, such as composite armour – the British Chobham armour, for example, comprising alternating layers of steel and ceramics – or

## PIERCING TANK ARMOUR

Two distinct types of anti-tank munitions keep the designers of tank armour constantly on their toes. The first is the shaped charge, as used in high-explosive anti-tank (HEAT) rounds, invented just prior to the Second World War. When a HEAT projectile hits it target, the explosive charge inside forces a stream of molten metal at high velocity through the armour and into the tank's interior. A subsequent development is the weapon known as the armour-piercing fin-stabilised discarding sabot (APFSDS), which uses kinetic energy rather than explosives to penetrate the target. The projectile is typically made of an extremely dense material, such as tungsten or depleted uranium. In the 1991 Gulf War, an American armour-piercing round of this type passed through two Iraqi tanks parked one behind the other.

explosive reactive armour to counteract the impact of charges, as on the Israeli Merkava tanks. Modern 'third-generation' tanks have even larger guns (120–125mm) that can fire while on the move. But they are also heavier, weighing on average some 55–60 tonnes, and enormously costly to produce.

## SLOPED ARMOUR

Because the trajectory of anti-tank missiles is horizontal, a simple way of improving the armour's resistance – effectively increasing its thickness – is to create a sloped surface. This also makes it more likely that the shell will ricochet off (the Germans learned this by bitter experience from the Soviet T-34 tank). All German tanks from 1942 onwards, such as the *Panther*, incorporated sloped armour. Other nations followed suit.

## MUSTARD GAS – 1917

# The advent of modern chemical warfare

**U**sed for the first time by the German army in 1917, mustard gas – so-called from its smell – was not the first poison gas to be deployed against troops. Nor was it the most deadly chemical used in the First World War. But the release of this colourless gas marked a turning point in the history of warfare.

They were artillery shells just like any other, except that a small yellow cross was printed on the casing. On the night of 12 July, 1917, the Germans fired 50,000 of them on the British trenches. The Tommies suspected nothing; even the sweet and spicy smell that arose was not unpleasant. But the sun rose to reveal a scene of carnage. Blinded soldiers staggered around, their faces and necks covered in blisters. Others had open sores all over their bodies, from which seeped a putrid yellowish fluid. Even morphine did little to alleviate the soldiers' dreadful suffering. The chemical agent responsible was dichlorethylsulphide. Its first use, near the Belgian city of Ypres, led to it being dubbed 'Yperite', but its smell earned it the name 'mustard gas'.

The deployment of Yperite marked a new chapter in the history of chemical warfare. The first gas attacks, involving chlorine and tear gas, had come in 1915 and since then crude respirators had been issued to troops. These consisted either of a gauze pad or a cartridge steeped in chemicals (such as washing soda) placed over the nose and mouth to trap and neutralise harmful agents. Such gasmasks protected the lungs, but offered soldiers scant protection against mustard gas, which also attacked exposed skin and mucous membranes. It could even penetrate clothing and, because it was oily and not very volatile, it could hang around in shell-holes and trenches for days. Yperite was not deadly if it did not get into the lungs, but it could still severely incapacitate soldiers and put them out of action for months.

**Pig's snout**
*The ARS 17 gasmask* (Appareil Respiratoire Spécial, '*Special Respiratory Apparatus*') *which was issued to French troops in 1918. The facepiece of rubberised cloth helped to protect the wearer against mustard gas.*

### A DEADLY CLOUD

**T**he first major use of gas in war took place on 22 April, 1915, at the village of Langemarck near Ypres. At 17.00 hours, when the wind was in the right direction, German forces released 168 tons of pressurised chlorine gas from some 5,730 cylinders. A greenish cloud spread out over a 4-mile stretch of the front and drifted towards Allied trenches manned by French Colonial troops from Africa and Martinique. Within minutes, hundreds of them were choking and retreating in panic. The German troops advanced, protected by makeshift masks, but failed to fully exploit the gas attack, which killed some 800–1,400 French soldiers and injured 2,000–3,000 more. The following day gas hit British lines.

**Gas!** *An illustration from* The Sphere *magazine shows a gas attack on British troops in April 1915, before gasmasks had become part of general kit.*

**First World War attack**
*German troops emerge from a cloud of phosgene gas during an attack on a British position in 1918. Gas was deployed as a shock tactic.*

## GASES AND THEIR EFFECTS

Chemical weapons are dispersed by being vapourised into tiny solid or liquid particles. Certain agents are merely incapacitating, being designed to put the enemy out of action without causing permanent damage. Others are lethal if untreated. Experts distinguish four different categories of agent: harassing agents, such as tear gases, which are largely non-toxic but act as sensory irritants; blister agents or vesicants (for example, mustard gas), which burn the skin and mucous membranes and can be fatal if they scar the lungs; choking agents, such as chlorine and phosgene, which cause damage (often fatal) to the lung-blood barrier; and lethal agents, which cause paralysis and death – these include nerve gases like sarin and the American binary agent VX.

## A long history

Chemical weapons were not a new idea. An Indian treatise on warfare dating from the 4th century BC explains how to make abrin, a deadly toxin similar to ricin. The period from antiquity to the late 19th century saw various other attempts to incapacitate enemies with noxious fumes, but there was no systematic development of chemical agents. The Hague Declaration of 1899 and the Hague Convention of 1907 expressly forbade the use of 'poison or poisonous weapons'.

Even so, in August 1914 the French were the first to use gas on the battlefield, when they released tear gas (ethyl bromoacetate) against

German positions. The Germans responded by firing irritants, followed by the choking agent chlorine and the even more potent phosgene. Thereafter, the production and deployment of chemical weapons escalated on both sides. Most of the compounds used – and their potential effects – were known about long before the outbreak of the First World War. Dichlorethylsulphide, for instance, had first been synthesised by the British chemist Frederick Guthrie in 1860.

The first of a new generation of deadly neurotoxins – agents that attack the central nervous system – was tabun, discovered in 1936, followed by sarin two years later. Sarin was manufactured by the Germans in the Second World War, but never deployed against the Allies for fear of mistakes and reprisals. Despite international conventions, many countries continued to stockpile chemical weapons in the Cold War. Some states even used them; Saddam Hussein's Iraqi forces, for instance, unleashed mustard gas on several occasions during the Iran-Iraq War of 1980–8.

**Alien appearance**
*US troops wearing M17A1 gasmasks during the 1991 Gulf War against Iraq.*

## A TRAGIC IRONY

Nobel chemistry laureate Fritz Haber, who was responsible for Germany's chemical weapons programme in the First World War, was of Jewish origin. He fled persecution after Hitler came to power in 1933, but not before he had had a hand in developing the pesticide Zyklon B in the 1920s, which was later used in the Nazi extermination camps.

# Exploring the unconscious

At a stroke, the Austrian doctor Sigmund Freud's discovery of the unconscious dispelled the ancient notion of the Psyche. Freud took the view that, where the mind was concerned, people were not always masters in their own houses. His ideas gave rise to a new kind of therapy which aimed to cure neurotic disorders through talking.

In late 1885 Sigmund Freud attended a series of lectures given by the French neurologist Jean-Martin Charcot at the Salpêtrière asylum in Paris. At the time, Charcot was using hypnosis and suggestion to treat patients with symptoms of hysteria. Though impressed, Freud soon realised that hypnosis provided only a temporary respite and so he began searching for a more effective method of treatment for nervous disorders.

On his return to Vienna, a female patient provided Freud with the inspiration for a new approach when she reacted to his constant questioning with a sharp retort: 'Don't interrupt me while I'm talking!' Consequently, Freud encouraged his patients to set aside all inhibition and talk through their problems in a 'free association' of ideas. The ultimate aim was the retrieval of unconscious memories, which Freud believed to be the source of all neuroses. In particular, he became convinced that shameful memories of sexual abuse in childhood were suppressed by the unconscious mind, only to manifest themselves later in life in neurotic behaviour. The technique was to form the bedrock of Freud's new therapeutic method of psychoanalysis.

## Oedipus, Eros and Thanatos

Freud formulated an elaborate theory of repressed sexual desire. At around the age of 4 or 5 years, he claimed, everyone – without exception – develops a deep sexual attraction to the parent of the opposite sex, a longing that is taboo. In 1896 he coined the term 'Oedipus Complex' for this mechanism, after the figure

**Mind experts**
*In 1909 Freud was invited to lecture at Clark University, Massachusetts, where this photograph with other leading psychoanalysts was taken. Freud is seated on the left; Carl Jung is seated on the right with Sandor Ferenczi standing behind him.*

**Sitting comfortably**
*The couch from Freud's consulting room in the Berggasse, Vienna, as re-created in the Freud Museum in London. Freud himself sat in the green armchair.*

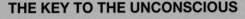

## THE KEY TO THE UNCONSCIOUS

In his work *The Interpretation of Dreams* (1900), Freud claimed that our dreams are manifestations of repressed desires. Although they often seem baffling at first, they can, he maintained, be deciphered like a puzzle, since even the most apparently absurd dream has its own internal logic. Freud regarded dreams as the most important route by which we gain an understanding of our unconscious minds.

in Greek mythology who kills his father and marries his mother. According to Freud, people are driven by a primal sexual pleasure principle (the 'Eros'), which civilised society demands that we sublimate to the 'reality principle'. Against the background of carnage in the First World War, Freud identified a counterpart to the Eros life instinct, a dark urge that impelled people to repeat experiences of suffering. He called this death drive 'Thanatos'.

## Conflicting impulses

In 1920 Freud laid down the definitive principles of psychoanalysis. Its keystone was the notion that the human mind forms the constant battleground for conflicts between three elements: the id, the ego and the super-ego. The id is the dark, inaccessible part of the human personality geared solely to indulging the pleasure principle. The ego represents the social being, which is subject to reason and common sense. The super-ego is a type of conscience that generates feelings of guilt. Neuroses are the result of conflict between these warring elements – a struggle that can only be resolved through psychoanalysis.

## A contentious issue

Recent discoveries of active molecules in the brain have prompted the development of drugs designed specifically to alleviate anxieties (for example, beta-blockers) or to stabilise mood (such as lithium for depression). Although drug treatment does not address the psychological causes of neuroses, it does enable patients to lead normal lives. Neurologists are now busy seeking explanations for mental disorders within the physiology of the brain. This work challenges Freud's central principle that mental illness is always the result of psychological conflict. There is even dissent among different schools of psychotherapy: psychoanalysis has been criticised for lack of clear results and the length of the treatment process, and has in large measure been supplanted by cognitive behavioural therapy (CBT). This more targeted approach focuses on resolving the debilitating aspects of a disorder without attempting to penetrate to the root cause (albeit at the risk of having it resurface in another guise).

What is beyond dispute is that Freud's ideas were born of a growing dissatisfaction among psychologists and philosophers in the late 19th and early 20th centuries at purely mechanistic conceptions of the human body and mind. Psychoanalysis took firm hold in the United States, where it is still a major form of therapy.

**Psychoanalysis in art**
*Entitled* Psychoanalysis and Morphology Meet, *this enigmatic painting by the Spanish surrealist artist Salvador Dali dates from 1939, the year after Dali was introduced to Freud by the novelist Stefan Zweig. The surrealists were concerned with states of mind and considered Freud a guiding light. Dali was inspired by psychoanalysis to devise a new creative approach as a way of committing his innermost hallucinations and deliriums to canvas. He called his method 'critical paranoia'.*

## NOT UNIVERSALLY VALID

Research by Polish anthropologist Bronislaw Malinowski among the Trobriand Islanders of New Guinea in 1915–18 discredited the idea that Freud's Oedipus Complex was a universal phenomenon. It was completely unknown in societies where children were raised by the whole community.

## SIGMUND FREUD – 1856 TO 1939

# Explorer of the human psyche

**S**igmund Freud devoted his whole life to delving into the unconscious mind, beginning with his own. He ascribed most of his major discoveries to this personal journey of self-awareness. By the time of his death just after the outbreak of the Second World War, psychoanalysis was a well-established discipline.

**Family man** *Sigmund Freud and his family in the garden of their house in Vienna (right). Seated in front of him is his sister-in-law Minna Bernays; his wife Martha is wearing the dark dress. Five of their children are in the picture – clockwise from bottom right: Ernst, Anna, Sophie, Oliver and Martin.*

**'My Golden Sigi'** *Freud at 16, with his mother Amalia. The following year, Sigmund went to university to study medicine, contrary to his father's wishes, who wanted him to follow in his footsteps as a cloth merchant.*

Sigmund Freud was born on 6 May, 1856, into a Jewish merchant family in Austria. When he was 5 years old his parents moved to Vienna, where he was to spend most of his life. His father was 40 when Sigmund was born, twice the age of his mother, and the powerful bond of love that he felt toward his mother as a young child later fed into his theory of the Oedipus Complex. From his father he inherited an insatiable thirst for knowledge. Sigmund was reading the Bible aged just 7 and the next year moved on to his first works of Shakespeare.

### Disciple of Darwin

Freud went into medicine not through any strong sense of calling, but because he was keen to study the brain scientifically. He spent years dissecting the nervous systems of fishes and read the works of Charles Darwin, whom he admired for basing his theory of evolution through natural selection on empirical observation. The young Freud was convinced that the human mind could be subjected to the same scientific rigour, free from religious prejudice and superstition. When Sigmund was 30 he married Martha Bernays, who bore him six children. His youngest daughter, Anna, would become his most faithful disciple.

### Transference and self-analysis

In 1886 Freud set up his first consulting room. As a succession of patients passed through his door, Freud was astonished at their attitude. They all seemed convinced that the therapist

**Devoted dad and daughter**
*Freud and Anna during a trip to the Dolomites in 1913. After training as a teacher, she became a pioneer of child psychotherapy.*

**Doctor Freud**
*The plaque from the door of Freud's consulting room at 19 Berggasse in Vienna.*

would supply all the answers to their problems. He called the feelings of attraction and admiration that they displayed towards him 'transference'. The entire aim of his therapeutic method was to get them to see that the solution lay not with him but deep within themselves.

Freud realised that his own anxieties were hampering his work as an analyst – that he was listening to his patients though the filter of his own questioning. So he subjected himself to a rigorous course of self-analysis, aimed at laying bare his own dreams and mainsprings of action as objectively as possible. His friend Wilhelm Fliess, to whom he wrote long letters, acted as his therapist. The experience confirmed Freud in the belief that nobody could become a psychoanalyst without first undergoing intensive analysis.

## A man alone

Freud's theories unsettled the scientific community. His former student Carl Gustav Jung openly challenged his theories on infant sexuality. Other colleagues criticised the fluid boundary he drew between the normal and the pathological and his assertion that everyone was neurotic in some way. Freud stuck to his guns undeterred. His Jewish upbringing in an intensely anti-Semitic Austria had taught him from an early age to face down hostility.

In 1934 Freud's works were burned by the Nazis. Forced to flee his native country, he settled in London in 1938. Some years before, he had been diagnosed with cancer of the jaw – he smoked up to 20 cigars a day, even smoking in the bath – and he finally succumbed to this illness on 23 September, 1939.

### DISCIPLES AND CRITICS

Alfred Adler, one of Freud's best-known followers, claimed that a person's social milieu is as important to their psychology as their inner life. Another pupil, Carl Gustav Jung, formulated the idea of the Collective Unconscious and the role of archetypes in shaping individual behaviour. Fellow-Austrian Melanie Klein pioneered child psychotherapy, seeking to explain adult problems through a child's relationship with its mother. Wilhelm Reich, who influenced many later German thinkers, tried to reconcile psychoanalysis and Marxism. The French thinker Jacques Lacan took a more orthodox Freudian line, but introduced two new insights: that a person's psychosexual development begins as early as 6-18 months of age, and that the unconscious is by no means chaotic, but rather is structured like a language.

# Motorways 1921

The world's first motorway was built in Berlin. In 1909 the German Automobile Club began investigating the feasibility of constructing a fast road to the west of the city. Work on the route, which was known as the AVUS (*Automobil Verkehrs und Übungsstraße*, 'motor traffic and test road'), was interrupted by the First World War and it opened on 21 September, 1921. Designed as both a regular road and a racing circuit, it set the pattern for all subsequent motorways: a dual carriageway separated by a central reservation to minimise the risk of collisions, with long fast straights and gentle sweeping bends.

Meanwhile, in 1914, a single carriageway road some 40 miles long had been built on Long Island, to enable workers to commute quickly into New York City. Nine years later an Italian civil engineer, Piero Puricelli, took the idea a stage further by creating the multi-lane junction, or interchange. Puricelli was also behind the development of the first long-distance motorway, the 'Autostrada' running the 28 miles (45km) from Milan to Varese, which opened in 1924.

**Work creation project** *Germany's Autobahn network, built under Hitler, reduced unemployment in the 1930s. Later, it would facilitate the rapid deployment of troops.*

The first motorway in Britain was the 8-mile Preston bypass, completed in 1958 and now part of the M6. The following year saw the opening of the first section of the long-distance M1 from Watford to Rugby, the road would eventually link London and Leeds. Tolls on motorways, a familiar feature in many countries including Italy and France, were long resisted in Britain but a toll road was opened in 2003 to relieve congestion on the M6 north of Birmingham. There are now more than 2,200 miles of motorway in Britain.

---

### LOCAL COLOUR

The sudden increase in the numbers of cars in the early years of the 20th century meant that planners had to come up with ways of ensuring safe traffic flow, especially in large cities. But the very first set of traffic signals pre-dated the motor car, being installed outside the Houses of Parliament in London on 10 December, 1868; it was a simple semaphore-like device, with red and green gas lanterns for night-time use. The modern electric traffic light was pioneered in Cleveland, Ohio, in 1914; it had just two colours, red and green. The first three-colour, four-way traffic light was introduced in Detroit in 1920. While these colours may be universal, possibly deriving from maritime signalling, the sequence and length of flashing varies from country to country.

---

Multi-level roads
*The first stack interchange, in which one main road crosses over another on a bridge with connector roads crossing on further levels, was the Four-Level Interchange in Los Angeles, built in 1949 (right).*

# The lie detector 1921

**Faking it?**
*Leonard Keeler advertised his polygraph by claiming that it could identify a genuine kiss. In 1935 he used it in earnest for the first time to question two criminals from Wisconsin. After establishing their guilt, they were duly sentenced.*

ROY POST · POSTOMETER

The lie detector, or polygraph, was invented by John Augustus Larson, a Canadian medical student, who unveiled his prototype machine in 1921. Based on William Moulton Marston's studies on the correlation between blood pressure and emotion, Larson's polygraph registered not only cardiovascular fluctuations, but also changes in breathing. The results looked convincing, but the diligent Larson never felt that the tests were sufficiently conclusive to market his machine as a bona-fide scientific instrument.

Larson's assistant, Leonard Keeler, was less scrupulous. In 1926 he built his own version, adding a monitor to detect levels of perspiration, and touted it around to various American political parties and law enforcement agencies. By the late 1930s the FBI and around a dozen police departments were using his lie detector. Most courts have since ruled polygraph evidence inadmissible, but almost a million tests are still conducted annually by American companies and other institutions.

## HOW TO FOOL A LIE DETECTOR

The American Aldrich Ames, who was exposed as having spied for the KGB between 1985 and 1994, successfully passed a CIA lie-detector test on two occasions. On the advice of his Russian handlers, he prepared beforehand with a good night's rest, then during the test remained calm and friendly towards the operator. Both times, the polygraph drew a blank.

# Puncture-repair kits 1921

In 1903, with more than a million cyclists on French roads, Louis Rustin opened a workshop in Paris to remould bike tyres. Rustin was a keen cyclist and knew only too well the problem of frequent punctures on badly maintained roads. He put his mind to producing a quick and effective solution to this recurrent problem. In 1908 he and his partner Jean Larroque began selling their first puncture repair kit, which comprised a strip of leather that could be glued inside the outer casing of the tyre. Thirteen years later, in 1921, he patented the puncture repair kit as we know it today.

Using the 'cold-cure' vulcanisation process, which kept rubber supple by steeping it in a solution of sulphur chloride, Rustin produced small rubber patches and adhesive to weld them to the inner tube. This simple cyclist's aid was an instant hit. By 1933 his company was turning out 28 million patches a month.

**Rubber Solution** 5gms ℮

**Quick fix**
*Puncture-repair kits are unchanged since Rustin's day: a tube of rubber solution, patches, a crayon to mark the hole and some chalk to rub on the glued patch to stop the inner tube sticking to the outer casing.*

## INTERIOR DESIGN
# All mod cons in the modern home

**W**ith the growth of a whole new manufacturing sector in the first decades of the 20th century, scientific and technological progress begin to have a direct impact on people's daily lives. Household goods and new labour-saving devices – vacuum cleaners, fridges, vegetable peelers – came thick and fast, taking some of the drudgery out of housework. The consumer society was born.

**Clothes iron**
*The Fulgator (above) was an electric iron made of wood and metal which had its heyday in the 1930s. Steam irons were introduced in the 1950s.*

**Home help**
*An early dishwasher, which came onto the market in 1926.*

Some domestic appliances made an appearance among the heavy engineering and arts and crafts on display at the Crystal Palace Great Exhibition of 1851 and at subsequent trade fairs. Yet the household goods market only really came into its own in Britain with the first Ideal Home Show, held at the Olympia exhibition centre in London in 1908. Sponsored by the *Daily Mail*, the event covered house construction, interior design, food and cookery, home furnishings and decor, and also included demonstrations of new products and competitions. The annual event grew steadily in popularity throughout the 1920s and 1930s. By 1957, as Britain finally began to emerge from its post-war austerity, the Ideal Home Show was attracting almost 1.5 million visitors.

---

### BREAKFAST CEREALS

**C**ereals have been valued for their nutrition since ancient times. As well as iron and other important mineral supplements, they are an important source of roughage and as such were known to be useful in combating digestive disorders. In 1898 Dr John Harvey Kellogg and his brother Will Keith began prescribing cereals at their sanatorium at Battle Creek in Michigan. Their rolled flakes of toasted grain proved so popular with patients that the brothers decided to launch full-scale commercial production. Thus, in 1906, the breakfast cereal 'Corn Flakes' was born.

# L'ART MÉNAGER

le N°. mensuel 4 Frs.

Octobre 1930

**Promoting modern living**
*As the consumer society continued to take root, so new magazines appeared aimed specifically at the housewife and home-maker. Both* Good Housekeeping *and* Ideal Home *were first published in the 1920s. This French publication (left) is from October 1930.*

NELLA CASA MODERNA

PAVIMENTI DI
**LINOLEUM**

SOCIETÀ DEL LINOLEUM · VIA M. MELLONI, 28 · MILANO

**Modern style**
*Advert for linoleum in the Italian architecture and design magazine* Domus *in 1928.*

## DURABLE AND EASY-CLEAN

Linoleum, or lino, is made by impregnating a backing cloth of burlap (jute) with a mixture of linseed oil and sawdust or cork dust. Pigments are then added to give it colour or a pattern. This waterproof, hard-wearing, hygienic flooring was invented in 1855 by the English manufacturer Frederick Walton, who patented it eight years later. Walton created the name from the Latin words *linum* ('flax') and *oleum* ('oil'). By 1869 his factory in Staines, West London, was turning out up to 3,000m² of lino a week, both for home use and for export to Europe and America.

## Lightening the housewife's load

The late 19th and early 20th centuries had seen the introduction of a plethora of labour-saving inventions that exploited electricity in the home. These included such devices as the electric iron (1882), kettle (1890), oven (1891), vacuum cleaner (1907), dishwashing machine (first sold in America in 1909), refrigerator (1913) and washing machine (1920). The first mass market for such goods was the United States, but after the First World War, as more women in Europe began to go out to work and affluence increased, with a corresponding rise in home ownership, so the demand for household goods expanded in Western Europe.

Significant demographic and social upheaval combined to change people's outlook and behaviour. Major pandemics like the Spanish flu of 1918–19 had depleted the workforce, and with better-paid factory work often a more attractive option than domestic service, home-making and domestic chores became a necessary preoccupation for most city-dwellers, including the growing middle class. This represented a radical break from the clear 'Upstairs-Downstairs' division of the Victorian and Edwardian eras.

## The hygienic kitchen

By the 1920s universal education and public health campaigns had made most people aware of the importance of cleanliness and hygiene in the home. Manufacturers played to this new preoccupation in their advertising. Most appliances unveiled at exhibitions were hailed as being great boons to hygienic living: vacuum cleaners, for example, were billed as sucking up harmful germs, while washing machines, it was claimed, disinfected the laundry.

In the kitchen wood was being replaced by more durable and easy-to-clean materials. Stainless-steel cutlery and pots and pans began to make an appearance from 1913 onwards, while heat-resistant glass Pyrex dishes from the USA – which first came on the market in 1915 – soon became a firm favourite with housewives. The modern American kitchen became the universal model, with brand-new electric appliances, purpose-built kitchen furniture and a shining white linoleum floor.

## Saving time and space

Interior design really began to change the way people lived. Furniture became more functional and better suited to small urban homes and flats. New designs for folding tables, beds and sofas and for stackable chairs helped to make

**Hot stuff** *An electric toaster from the 1910s (left), with a two-pin Bakelite plug for fitting into a light socket. The electric toaster became possible following the invention – by two American metallurgists in 1905 – of high-resistance nickel-chromium wire for the heating elements.*

**Design icon**
*In 1927 the Dutch architect Mart Stam produced his elegant S33 cantilever chair. Made from leather and chromed steel tubing, the chair is light and stackable and rapidly became a design classic.*

### POPULAR APPLIANCE

**B**efore the advent of electricity, slices of bread were toasted by holding them in front of a fire on a long toasting-fork or in a metal frame. In 1909 the American technician Frank Shailor patented the first electric toaster, which was sold in New York by the General Electric Company as the GE D-12. It was mounted on a ceramic base (some models were decorated with floral patterns) and comprised four heating elements made from nichrome alloy wire, which glowed red-hot. This first model could only toast one side of the bread at a time and did not eject the slices when they were done. The automatic pop-up toaster, in which the heating element was on a timer, was perfected by the American Charles Strite in 1919. Launched in 1925 as the Toastmaster, Strite's machine browned both sides of the bread simultaneously. Since then, the toaster has gone from strength to strength. Sophisticated modern models can defrost frozen bread, heat up croissants and collect stray crumbs in a sliding crumb-tray.

the most of limited space. To the same end, some pieces of furniture even performed multiple roles, such as ottomans that doubled as a seat and as a blanket box.

But while some manufacturers placed a premium on space-saving design, others chose to stress how much time their products saved the hard-pressed housewife. A welter of gadgets to make housework less of a drudge began to appear on the market, such as vegetable slicers (moulis), salad spinners, shoe-cleaning machines and egg-boilers. The same period saw the spread of convenience foods. Established products such as Oxo cubes from the Liebig Company (which had been in existence since 1865) and Maggi instant soups (1883) rubbed shoulders on kitchen shelves with newer products like the malt drink Ovaltine (1904), the yeast-extract spread Marmite (1902) and tea bags (1919).

**Flavour in an instant** *The social trend away from domestic service meant that time-consuming kitchen tasks, such as making meat stock from bones, went by the board. Factory-made beef-extract products such as Oxo and Bovril filled the gap. This 1925 Italian poster (left) advertises the convenience food products of Liebig's Extract of Meat Company (Lemco).*

## IN SEARCH OF THE PERFECT CUP OF COFFEE

The idea behind the cafetière – now found in millions of homes around the world – is to make a jug of coffee without running the risk of boiling the liquid, which ruins the flavour. It does this by allowing piping-hot (but not boiling) water to gently percolate through a filter holding the ground beans. Frenchman Jean Baptiste de Belloy was the first person to invent a cafetière-type device in 1800. It involved two containers separated by a small compartment containing the ground coffee. Water was poured in at the top and slowly passed through the coffee, before dripping into the lower flask. The first 'Cona' vacuum cafetière, comprising two glass globes, appeared in 1825. One globe was filled with water, which was heated by a small burner; as the steam rose, it condensed through a tube into the second globe, which held the ground coffee. When all the water had evaporated

and dripped through, the flame was snuffed out; as the first glass globe cooled and the air in it contracted, this created a vacuum that sucked the coffee back through the tube (which had a filter in it). Further innovations included the introduction of filter papers by Melitta Bentz in 1908, and the invention of the plunger-type cafetière by the Italian Attilio Calimani in 1929. The definitive modern espresso machine was unveiled by Achille Gaggia in 1948.

**Pick-me-up**
*A 'Cona' coffee percolator from the 1950s, made by Kirby, Beard & Co.*

# Soaring like a bird

The English aviation pioneer Sir George Cayley had come up with a working design for a glider over a century earlier, but the sport of gliding (also called 'sailplaning') only really took off after the First World War. The first true gliders had canvas-covered wooden fuselages and wings, but over time evolved rapidly with the introduction of new materials such as fibreglass, carbon fibre and Kevlar.

**The glider king**
*Sailplane pioneer Otto Lilienthal in flight in 1893. He died in a glider accident three years later; his final words were 'sacrifices have to be made'.*

**Learning from the birds**
*The French sailor Jean-Marie Le Bris was inspired by watching seabirds to design a machine called the* Albatross, *which he built with the help of the French Navy. It had flapping wings with a span of 18 metres and made three short hops near Brest in 1868.*

The great catalyst to the development of gliding was the Treaty of Versailles in 1919. In a move intended to prevent the re-emergence of Germany as an aggressive military power, the treaty strictly limited the nation's ability to manufacture single-seater powered aircraft. As a result, German air enthusiasts turned to sailplanes and the country became the birthplace of modern gliding. The first gliding competition was organised in 1920 at the Wasserkuppe, a hill in the Rhön Mountains near the town of Fulda in Hesse. It was there, two years later, that a pilot called Arthur Martens was the first to use an updraft rising off the slope of the Wasserkuppe, enabling him to stay airborne for a significant period.

Gliding was also gaining in popularity elsewhere. The first French competition for unpowered aircraft was held near Clermont-Ferrand in 1922. Britain's inaugural gliding club was founded in Kent in 1930, but two

**BUNGEE LAUNCHES**

One method of launching a glider that is no longer widely used is the bungee launch, invented by the German gliding pioneer Dr Wolfgang Klemperer. This involved positioning the glider, facing into the wind, at the top of a long incline. Crewmen would then run down the slope holding elasticated bungee cords attached to the glider's nose, while others held on to its tail. On a signal, the glider was released, and as it shot forward and got airborne, the pilot disengaged the bungees.

decades earlier Eric Gordon England had gained the distinction of being the first man to make a soaring flight when he rose, albeit briefly, to a height of 30 metres over Amberley Mount on the South Downs in Sussex.

## Mastering the art of soaring

The earliest would-be pilots of heavier-than-air craft took to the air either by hurling themselves off high vantage points or by using kites to harness updrafts. The inspiration for their conviction that unpowered flight was possible came from observing birds riding air currents without flapping their wings. Many of these early, more or less controlled descents were undertaken with an eye on the greater prize of achieving powered flight, and could not be regarded as gliding, which involves not just staying aloft but gaining height.

People only gradually began to understand the nature of thermals during the 20th century. When the first aviators discovered the existence of rising air currents, it rekindled the age-old human dream of soaring like a bird. Meteorologists distinguish between two different types of rising air current: thermals, which come about through convection (as bubbles of warm air heated by the Sun rise up into the colder surrounding atmosphere), and wave lift, which is caused by wind being deflected upwards by high ground. This second type gave rise to the phenomenon of wave flying and has enabled glider pilots to set flight endurance records of almost 60 hours – in 1952, for instance, a French pilot flew for 56 hours and 15 minutes. Such record attempts were discontinued after several fatal accidents caused by pilots falling asleep.

The reason why some birds can keep soaring upwards, despite being subject to the force of gravity (like every other body on

Earth), is that the air mass in which they circle is rising faster than the rate of descent. Others take advantage of the lift produced by updrafts from hillsides and ridges. Likewise, gliders are designed to minimise their rate of descent.

## Training tomorrow's fighter pilots

The records set by the German pilots Arthur Martens and Friedrich Hentzen – one to three hours of flying – seemed like extraordinary exploits in August 1922. But just nine years later Günter Groenhoff flew 170 miles (270km) on a storm front from Munich to Kadan in Czechoslovakia. A gliding school was established on the Wasserkuppe in 1925 and leading aeronautical engineers like Willy Messerschmitt and Alexander Lippisch tested their early designs there.

**Novel start**
*Trying to launch a glider from a moving car in Bagatelle Park near Paris in 1910.*

**Bird-like design**
*A Grandin glider taking off at a sailplane meeting near Cherbourg in August 1923.*

accounts for almost one-third of all qualified glider pilots in the world, and the three main manufacturers of gliders are all based there.

## Constant refinement

The first gliders were made of wooden or tubular metal frames covered with canvas. From the 1970s, these were superseded first by machines made from plastics then later of composite materials such as fibreglass, carbon fibre or Kevlar. Gliders took on ever more elegant and aerodynamically efficient shapes, while the wingspan increased enormously, making this new generation of single-seat and two-seater gliders capable of far longer flights than earlier craft. Advances in electronics over the same period made flight instruments more

**Young eagles** *Members of the Hitler Youth Flight Corps (above) study the principles of flight with the aid of a model glider.*

When the Nazis came to power in 1933, glider training was brought under state control. The Wasserkuppe became the nursery for the country's resurgent air force, the Luftwaffe, which Hitler unveiled in 1935. Many future Second World War pilots, including Erich Hartmann, the top-scoring fighter ace of all time, learned to fly on this programme. In other countries, gliding remained something of a marginal sport for amateurs. Germany still

**On top of the world**
*A favourite, if dangerous, pastime among experienced glider pilots is mountain soaring (above), which enables them to fly huge distances. The German pilot Klaus Ohlmann holds the current record with his flight of 3,009km (1,870 miles) over the High Andes.*

---

### ASSAULT GLIDERS

Assault gliders were developed in Germany at the start of the Second World War and were used by airborne German forces to capture the Belgian fortress of Eben-Emael in May 1940. Cheap and robust, they were towed by transport aircraft close to the drop zone, then left to glide a short distance to the target. Most were designed to ferry troops, but larger gliders could carry jeeps and even light tanks. The Allies deployed glider-borne forces to capture forward positions in the D-Day landings. After the war, the invention of the helicopter made assault gliders obsolete.

---

**Gliding into battle**
*Troop-carrying* Horsa *gliders fly over the Normandy beaches as part of the Allied airborne assault on German-occupied France in June 1944. They were towed into position by C-47* Dakota *transport aircraft.*

**Freedom of the skies**
*If meteorological conditions are favourable, modern gliders (below) can cover 300 miles or more. But when weather conditions deteriorate, non-powered gliders are sometimes forced to 'land out', that is to ditch in a suitably flat field.*

sensitive and compact. Launch methods had also kept pace, as aero-towing replaced winch or bungee launches. Some modern gliders are fitted with light, low-powered engines (some even retract into the fuselage) so they can take off without a tow.

The principal measure of performance in modern gliders is the glide ratio, which measures distance travelled in terms of height lost. Thus, a glider with a glide ratio of 40:1 can travel for 40km while losing 1,000 metres of altitude. Gliders with a 15-metre wingspan, the most common type today, have glide ratios of 40 or 50:1 (the best pre-war designs had ratios of 25:1). The most advanced gliders, with 30-metre wingspans, have glide ratios of more than 70:1. (As a comparison, most airliners have a ratio of between 8 and 11:1.)

There are estimated to be around 120,000 active glider pilots worldwide. The sport's great appeal lies in the quietness of the flight and sense of freedom. With the cockpit enclosed under a Plexiglas canopy, glider flying is no longer the gruelling experience it once was, but it still demands great concentration and expert knowledge of aerodynamics and meteorology.

## HIGH FLYERS

The absolute altitude record for gliders is currently held by the multi-millionaire American businessman and adventurer Steve Fossett, who was killed in 2007 while flying over the Nevada Desert. In 2006 he and his co-pilot Einar Enevoldson flew into the stratosphere to a height of 15,460 metres (50,727ft). Prior to that flight, the sailplane altitude record was held by another American, Robert Harris, who reached 14,938 metres (49,009ft) in 1986. This type of record can only be achieved by wave flying, where the pilot – who must wear a fully pressurised flying suit – gains enormous lift by soaring above high mountain ranges.

# SUPERMARKETS
# One-stop food shopping

Forty years before Elvis, on 6 September, 1916, a buzz went round Memphis, Tennessee. At 79 Jefferson Street, a store named 'Piggly Wiggly' opened its doors for the first time to reveal a new concept in shopping, the brainchild of wholesale grocer Clarence Saunders. Frustrated by the inefficiency of traditional shops, Saunders set out to revolutionise the food retail sector. Shopping would never be the same again.

**Back to the future**
*A self-service grocery in Milwaukee in 1912. The first grocery to offer partial self-service opened in 1896 in New York; all its goods were price-tagged and it offered a money-back guarantee to dissatisfied customers.*

Clarence Saunders began his crusade by introducing the concept of 'cash and carry': retailers bought their goods directly from his wholesale warehouse, paid cash, then took their purchases away themselves rather than having them delivered. The end result was that everybody profited. The success of the formula encouraged him to extend the concept to the retail sector by opening a pilot self-service store. The old model of a grocer behind the counter, with queues of customers waiting to be served, would become a thing of the past almost at a stroke. In 'Piggly Wiggly' stores,

goods were pre-packed, clearly labelled and openly displayed on shelves for shoppers to help themselves. The self-service format made food shopping quicker and more efficient. Customers were in and out much faster and they paid less, as the smaller workforce meant lower overheads for the store. What's more, the retailer's profit was boosted by the fact that people were more likely to make impulse buys when they had the freedom to walk around and choose goods for themselves.

The measures that Saunders introduced to prevent shoplifting brought further refinements to his store. He began by putting in two doors, one for entry only and the other for the exit. He then installed turnstiles to channel the flow of people leaving and put the tills in front of the exit. Within just a few months, thefts fell from 6 per cent to just 0.75 per cent of total turnover. All the key elements of the self-service supermarket were now in place.

## COOL-CHAIN DISTRIBUTION

Despite the fact that the Frigidaire fridge had been on sale in the USA since 1919, it took until 1930 and the introduction of the freon for the refrigeration industry to really get going. From the 1940s cold-stores, refrigerated lorries and supermarkets chillers meant that deep-frozen foods could be sold in tip-top condition.

## THE RULES OF RETAILING

**B**ernardo Trujillo was a marketing guru employed by the American National Cash Register Company (NCR) of Dayton, Ohio, to train shopkeepers around the world in self-service sales techniques and high-volume retailing. He offered the following advice on how to attract customers: 'For a product to sell well, it needs to look like Brigitte Bardot. It must be beautiful, well presented and readily available'. Several factors played a part in the new marketing: the length and height of the shelves, the impact of signage, good lighting, an unimpeded flow of shoppers down the aisles and even the use of piped music in stores. Trujillo's seminars were legendary in laying the foundations of modern retailing.

**A step too far**
*Clarence Saunders in one of his fully automated grocery stores. At these 'Keedoozle' outlets (from 'Key Does All') shoppers were given a keyboard, which they used to select items. Their purchases were tallied on a till roll (above right) and the goods delivered on a conveyor belt. Keedoozles were not a success; only three stores were ever built.*

In the 1930s, as the Great Depression took hold, the entrepreneur Michael J Cullen took the self-service concept a stage further. Buying up an empty garage in Queens, a densely populated district of New York, he turned it into a huge no-frills store, where people could buy goods at wholesale prices. Customers flocked to the King Kullen Grocery Store in their droves, attracted by the range of products on offer and the low prices. By 1936, Cullen had a chain of 17 supermarkets with an annual turnover of 6 million dollars.

## Europe follows suit

It was only in the 1940s and later that self-service shopping came to Europe. London's first self-service grocery opened in 1942; its first supermarket nine years later. Retailers in France (Paris, 1948), Switzerland (Zurich, 1952), and Germany (Cologne, 1958) then caught the self-service bandwagon.

The siting of the first supermarkets was carefully planned, taking into account such factors as ease of axis and proximity to post offices and schools. From the outset, the aim of supermarkets was to offer housewives everything they needed to run a home under

**At your service**
*Cashiers at the checkouts of a 'Piggly Wiggly' supermarket in Encino, Los Angeles, in 1962. The origin of the name remains a mystery.*

**Aisle of plenty**
*German photographer Andreas Gursky created this image (right) of a hypermarket interior in 2001. One of a pair entitled 99 Cent, the images each measure 4 by 2 metres, conveyimg by their sheer size something of the vastness of these huge retail halls.*

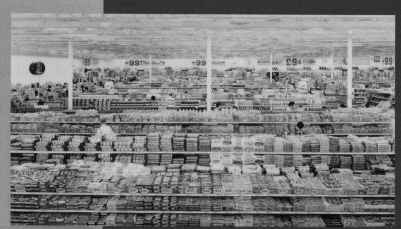

## SHOPPING TROLLEYS

The prototype of the shopping trolley was invented by Nathan Goldman in 1936, and comprised two wire-mesh baskets fitted on top of a frame with castors. Ten years later Orla E Watson introduced shopping carts fitted with loose gates at the back that enabled them to be pushed together when empty. By the 1950s shopping trolleys looked much like they do today, but on average they were five times smaller than the largest trolleys currently in use.

### Ongoing innovation

In the 1960s and 1970s, some continental retailers tried some offbeat innovations to expand market share. In 1963 the first 'drive-in' supermarket was opened in Bavaria, but it was not as advanced as it sounds. It entailed customers selecting their goods in advance from a catalogue, driving to the store and handing in their list, then waiting while staff fulfilled their order. The idea never really caught on. Other novelties included a floating supermarket in Rotterdam and tent supermarkets set up on camping sites in Denmark by the firm Irma.

The main trend was to build bigger and bigger supermarkets, with free car parks for their shoppers. As automobile ownership increased throughout the Western world, large-volume retailers of groceries opted to build so-called 'hypermarkets', accessible only by car, on the outskirts of towns and cities where large tracts of land for development could be bought for a fraction of the price of a town centre site. Latterly this trend has attracted much criticism for turning town centres into 'ghost towns', as reduced trade has forced smaller specialist shops out of business and large chains have relocated to purpose-built shopping malls, another US-inspired phenomenon. Ecological concerns have prompted many planning authorities to call a halt to new developments of this kind.

Recent technological in-store innovations include touch-screen self-service tills. These help to reduce queues and lower the retailer's wages bill. Some supermarkets operate a 'quick check' service, where shoppers use a hand-held scanner to tally up their bill as they go around the store; this also saves time at the checkout. Also, in a high-tech throwback to the pre-supermarket era when grocers and other individual tradesmen would deliver orders to the customer's home, the Internet today enables people to do their supermarket shopping online and have the goods brought direct to their door.

one roof. Jack Cohen, who began trading from a market stall in 1919, opened the first Tesco supermarket in 1948. His famous motto was 'Pile it high and sell it cheap'. By 2010 Tesco had grown to become the third largest retailer by revenue on the planet, after the American giant Wal-Mart and the French chain Carrefour. It was Carrefour that pioneered the sale of fuel from its forecourts at a discount over traditional petrol stations, a practice since widely copied by other supermarkets.

*Supermarket shopper*
*A life-like resin sculpture (above), made in 1970 by the hyperrealist artist Duane Hanson, satirises the American tendency to overconsume.*

# The bulldozer 1923

The history of the bulldozer began in 1923 at Morrowville, Kansas, on the site of a pipeline being built by the Sinclair Oil Company. A young local farmer named James Cummings noticed that, while the trenches for burying sections of the pipeline were excavated by steam-shovel, the job of backfilling was still done with a pair of mules dragging a board. He approached the oil company suggesting a machine might perform this task and was encouraged to develop his idea.

With the help of local draughtsman Earl McLeod, Cummings scoured scrapyards in the area for parts and built an earth-moving machine. They based it on a steam-powered caterpillar farm tractor manufactured in 1904 by Benjamin Holt (who went on to found the Caterpillar Tractor Company). To this basic vehicle, they attached two tools: a wide earth-moving blade at the front and a ripper, a claw-like device, on the back to break up surface rock or impacted soil. The bulldozer was an instant success. Sinclair contracted Cummings and McLeod to backfill the entire length of the pipeline. Thereafter, more powerful bulldozers were developed for building sites and road construction projects, while small versions were built for work on restricted-access sites.

**Heavy lifter** *A heavy-duty earth-moving vehicle, running on massive tyres rather than caterpillar tracks, in a quarry.*

### ARMOURED BULLDOZERS

Huge bulldozers fitted with armour-plating, bullet-proof cabin glass and metal grilles covering the windows to protect the driver from attack are used by several armed forces for combat operations. These vehicles – notably the massive D9 manufactured by the Caterpillar Company – have become controversial weapons of war. In both the 1991 Gulf War and the invasion of Iraq in 2003, US forces were accused of using them to bury enemy conscripts alive in their trenches. The same vehicle has been widely used by the Israeli Defence Forces to destroy the homes of Palestinians accused of terrorist offences. US peace campaigner Rachel Corrie was crushed by an Israeli D9 while protesting against these operations in 2003.

# Paper hankies 1924

Legend has it that Ernst Mahler, the head of the Kimberly-Clark paper company, got the idea for the paper hankie when he blew his nose one day on a cellulose wipe designed for removing make-up. It struck him that this was more hygienic than carrying around a traditional cloth handkerchief full of germs and he realised there could be a market for disposable tissues.

The 'Celluwipe', as the product was first called, also dispensed with the chore of laundering handkerchiefs. Rebranded as 'Kleenex', it went from strength to strength. More recently, in view of concerns about the sustainability of throwaway goods, Kleenex have created tissues of biodegradable natural fibres with Forest Stewardship Council certification.

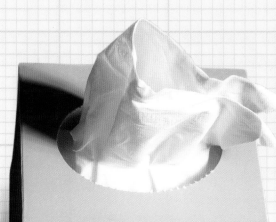

**Handy hanky**
*The tissue box with a perforated tear-out lid was invented by Kimberly-Clark in 1928.*

# Unlocking the brain's secrets

Wh_at happens in the brain when someone is thinking, sleeping, or focusing on an object? The first person to record activity within the human brain in response to various stimuli was Hans Berger, a German psychiatrist. Electroencephalography, as he termed his procedure, would revolutionise our understanding of mental processes.

People had known of the existence of electrical currents in the brain since 1875 when Richard Caton, a Liverpool surgeon and medical school lecturer, discovered brain signals by measuring electrical potentials on the exposed cortex of rabbits and monkeys. But the true nature of these impulses was not revealed until 6 July, 1924, when the German psychiatrist Hans Berger identified for the first time a cortical current on the surface of the brain in a patient at the psychiatric clinic that he ran in Jena.

## Faint stirrings

The apparatus Berger employed, which he dubbed an 'electroencephalograph', was fairly primitive. Silver foil electrodes were attached to the skin of the patient's scalp by means of an elastic bandage; wires from these electrodes ran to an amplifier and a string galvanometer. Any oscillations detected by the galvanometer were to be reflected in the movement of a needle and recorded graphically as a trace on a long moving strip of paper. To achieve a visual recording was a real challenge, since the electrical currents coursing through the brain are extremely weak, measuring barely one ten-thousandth of a volt.

As the anxious researcher looked on, the recording needle began to flicker slightly, tracing a line made up of hundreds of peaks and troughs. Faint though the trace was, Berger was delighted with the result, having expended great effort and gone through many

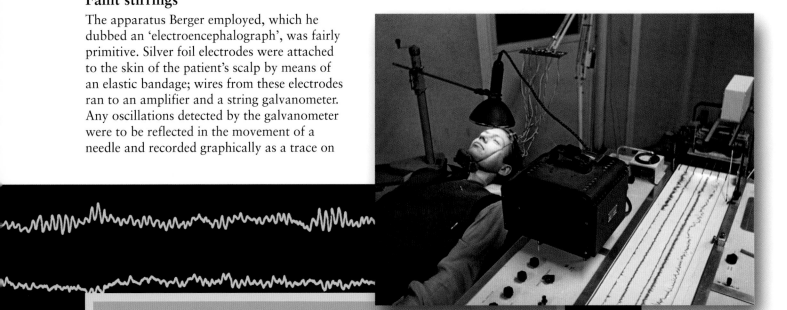

**Hooked up**
*A person undergoing an EEG test in the 1950s (above). Around this time, the procedure was first used to look for signs of epilepsy in patients.*

### STATES OF THE BRAIN

In his 1929 paper Berger described two very distinct types of brain wave that he had identified: 'The electroencephalogram represents a continuous curve with continuous oscillations in which ... one can distinguish larger first-order waves with an average duration of 90 milliseconds and smaller second-order waves of an average duration of 35 milliseconds ...' The first-order (alpha) waves are seen in states of relaxation, especially when the eyes are closed, while beta waves are associated with mental concentration and with startle reactions. Later neurophysiologists also identified low-frequency delta waves, which occur in deep sleep (the phase when people are most likely to talk or sleepwalk). After 60–70 minutes of sleep, these are superseded by a rapid, low-voltage EEG, accompanied by rapid-eye movement (REM) and low muscle tone. This is so-called 'paradoxical sleep' (also known as the 'third state' of brain functioning, after alpha and beta waves), during which sleepers customarily dream.

inconclusive trials to get to this point. Even so, to be sure that the electroencephalogram really was recording an electrical brain phenomenon, he continued with the experiments for five more years in the small laboratory in the basement of his clinic, using his patients, his son Klaus and even himself as guinea pigs for study. Berger always conducted his testing on Sundays when the electrical devices in the hospital were switched off, so they could not interfere with the registration of the tiny impulses on the encephalograms.

## From scepticism to recognition

By 1929 Berger had identified two types of oscillation, which he termed 'alpha' and 'beta' waves, and was sufficiently convinced of his findings to publish his first paper that July, entitled 'On the electroencephalogram in man'. But his results were greeted with scepticism by his peers, despite the fact that they confirmed similar readings of brain waves in dogs, made in 1912 by the Russian physiologist Vladimir Pravdich-Neminski. Professional snobbery played a part in the rejection; the medical establishment looked down on Berger and his work because he was not a neurophysiologist.

In 1934 Edgar Adrian, a neurophysiologist at Cambridge who had recently been awarded the Nobel prize for physiology, also recorded the alpha wave, employing an improved EEG apparatus that used copper-gauze electrodes wrapped in saline-soaked lint. In a paper describing this research, published jointly with his assistant B H C Matthews, Adrian cited Berger's seminal work, acknowledging his influence. In a further mark of recognition, Adrian invited Berger to be co-president of the first international symposium on electrical activity in the nervous system, held in Paris in 1937. Despite this belated recognition Berger, who was prone to bouts of deep depression, took his own life in 1941 in the clinic where he had done his groundbreaking experiments.

Since the 1950s, the EEG has become the standard method of studying the brain and diagnosing its disorders. It can reveal tumours, vascular deformities and infections in brain tissue, and is a vital first step before more thorough investigation using an MRI or CAT scanner. It has also helped in the identification and investigation of different forms of epilepsy, a disease that is related to chronic or temporary hyperactivity in certain regions of the brain.

**Painless procedure**
*Taking an EEG is a non-invasive procedure that involves attaching electrodes to the scalp – usually 16–20, far fewer than in the photograph above. The test generally lasts no more than an hour, during which the patient is subjected to visual stimuli that stimulate particular types of brain activity.*

## VIBRANT VIENNA
# A city old and new

The collapse of the Habsburg Empire following defeat in the First World War had a dramatic effect on the Austrian capital. As the old imperial forces made way reluctantly for socialist 'Red Vienna', the city became a magnet for the avant-garde of the Modernist movement. A visitor to Vienna in the 1920s would have been struck on the one hand by the city as an historical treasure trove, on the other by its dynamic new spirit.

**The march of time**

*A postcard from the early 20th century (below) shows the modern world encroaching on medieval Vienna. Motor cars and omnibuses mingle with the crowds and horse-drawn carriages on the streets around Stefansplatz, the central square in the heart of the Austrian capital dominated by St Stephen's Cathedral.*

In the first decades of the 20th century, the capital of the defunct Austro-Hungarian Empire resembled a vast museum. In 1919, in the aftershock of the First World War, Austria had shaken off its imperial past and become a republic, but it was business as usual at the Hofburg, once the emperor's palace, now the official residence of the president. Musically the city continued to resound to the symphonies, operas and chamber music of the great Baroque and Classical composers who at various times had lived there, among them Gluck, Haydn, Mozart, Beethoven, Schubert and Brahms. This noble tradition was upheld by more recent composers, such as Anton Bruckner and Gustav Mahler. And the ever-popular waltzes of the Strausses (Johann the Elder and Johann the Younger) could be heard every Sunday played by numerous bands on the Prater, the city's great leisure and amusement park along the River Danube.

### Birthplace of psychoanalysis

Despite the nostalgic atmosphere, modernity had not passed Vienna by. One of the most radical thinkers of the age, the psychoanalyst Sigmund Freud, regularly held court at the Café Landtmann, a favourite haunt of the intellectual élite, surrounded by his circle of followers, journalists and politicians. Freud's theories were not just the talk of the town but of the whole world at this time. He claimed to have found a cure for the age-old curse of melancholy, tracing it to unconscious traumas and conflicts that stemmed from people's experiences in early childhood. Freud's psychoanalytic therapy came too late for Crown Prince Rudolf von Habsburg, the liberal-minded heir to the throne who took his own life in 1889 while in the grip of a deep depression. If he had lived, might the empire have been saved? Shorn of Hungary and its extensive possessions in the east, Austria was now a mere shadow of its former self.

### Cradle of modernism

A leading light of the Viennese intellectual scene had been the physicist and philosopher Ernst Mach, who died in 1916. It was Mach who ascertained that the inertia of a body is a function of the relationship between that body and the rest of the Universe. He claimed that there was no such thing as absolute time or space, that these concepts could only be relative. Albert Einstein later wrote that it was Mach's ideas that inspired him to formulate the theory of relativity. Mach was also famous for establishing new measurements for expressing the velocity of a body in relation to the speed of sound, now commonly used in aviation and space travel as Mach 1, Mach 2 and so on.

Another Viennese thinker was about to turn Western philosophy on its head. Ludwig Wittgenstein (1889–1951) published his one and only book, *Tractatus logico-philosophicus*, in 1921. It made a profound impression on the

WIEN I.
Stefansplatz,

**Functional showcase**
*The Scheu House (left), in the Hietzing district of Vienna, was a private dwelling built in 1912 by the architect Adolf Loos. Devoid of unnecessary showy ornament, it demonstrated his love of simple design.*

WIENER CAFE: DER LITTERAT.

English philosopher Bertrand Russell, the foremost mathematician of the period, and it revolutionised mathematics, formal logic, language and philosophy. Wittgenstein expounded the view that words, like numbers, are mere postulates rather than facts and that their meaning is therefore uncertain. He also claimed that truths obtained by logical deduction were tautologous – that is, they were simply circular reasoning. The radical deconstruction of thought and language that Wittgenstein set in train resonated down the 20th century in the works of others, including the French philosopher of language Jacques Derrida. Genius seemed to run in the family: Wittgenstein's second cousin was Friedrich Hayek, who went on to become one of the leading economists of the modern age.

## Shock of the new

Even before the First World War, Modernism was making itself felt in Vienna. In the midst of the Baroque churches and ornate façades on the Michaelerplatz, facing the Imperial Hofburg Palace, an extraordinary new building was completed in 1911. Striking in its stark simplicity and lack of embellishment, the Goldman and Salatsch Building formed a glaring contrast to its surroundings. It also

outraged the arch-conservative Emperor Franz Josef II. It was popularly known as the 'Looshaus' after the architect Adolf Loos, its creator. Loos believed that buildings should first and foremost be functional. He was also a firm devotee of the new reinforced concrete. In the post-war period his new architectural idiom, made all the more attractive by the speed and low cost of construction, was eagerly seized upon by Austria's ruling Social Democrats for their ambitious programme of slum clearance and the construction of new apartment blocks for the city's workers.

**Arts and crafts**
*The Wiener Werkstätte aimed to unite the fine and applied arts in creating functional objects of beauty. From 1907 it issued postcards by leading Viennese artists of the day, including this one by Moritz Jung.*

### HIGH PRIEST OF THE FREE MARKET

After Adam Smith in the 18th century, Friedrich Hayek (1899–1992) was, along with his British contemporary John Maynard Keynes, the economist who had most impact on market economics. He thought Soviet-style economies were doomed to failure. In contrast to the command economy advocated by Marx, Hayek believed that economic systems cannot be micro-managed by the state. Only the spread of information, made possible by the growth of the popular press and radio broadcasting, can influence their evolution. Hayek's huge body of work, which also addressed questions of jurisprudence and political theory, won him the Nobel prize for economics in 1974.

The new modernist style could also be seen in Vienna's coffee houses, where Cubist-inspired tea and coffee sets made of stainless steel began to appear, alongside crockery designed by Josef Hoffmann and Koloman Moser who had founded the *Wiener Werkstätte* ('Vienna Workshops') in 1903. This design studio was to have a huge influence on consumer goods worldwide, imposing an aesthetic of clean lines and absence of fussy ornamentation. Culinary delights in the coffee houses of the Austrian capital, meanwhile, included Sachertorte, a rich chocolate and apricot jam cake that was a speciality of the Sacher café, and Schlagobers, a dark coffee topped with whipped cream, which became popular throughout Europe.

## Viennese atonality

In the 1920s three young composers of the so-called 'Second Viennese School' – Arnold Schoenberg, Anton Webern and Alban Berg – began to develop a new form of music that

**Clear craftsmanship**
*An armchair designed by Josef Hoffmann in 1908. The chair soon earned the nickname* Sitzmaschine *('sitting machine') from the nuts and bolts holding it together, which the designer made no attempt to conceal.*

## URBAN VISIONARY

Otto Wagner, a pioneering architect of the Modernist movement, left his mark on Vienna. Originally an exponent of the neo-Renaissance style, from 1894 – the year he became Professor of Architecture at the Academy of Fine Arts – Wagner embraced Art Nouveau (*Jugendstil*), which was then fashionable throughout Europe. A classic example of Wagner's work is the entrance to the Karlsplatz underground station, with its repeated floral motifs. Another of his buildings is the Austrian Postal Savings Office, a reinforced concrete structure that marked a new move towards functionalism. Wagner was interested in urban planning and in 1890 drafted an ambitious plan for the redevelopment of Vienna, including a complex network of railways, stations and viaducts.

**Jugendstil jewel**
*Wagner's Karlsplatz station building was saved from demolition in 1981 and now houses temporary exhibitions devoted to the architect's work.*

KARLSPLATZ

**Serial composers** *A caricature sketch made by Ehrt Rainer in 2004 shows, from left to right, Arnold Schoenberg, Alban Berg, Anton Webern and Hanns Eisler of the Second Viennese School.*

made a radical break with the clear harmonies and melodic structure of the classical canon. Known as 'atonal music' or 'serialism', their new compositional technique used all 12 tones of the chromatic scale equally. In a bourgeois Vienna still in love with waltzes and comic light operettas, their experimental approach fell on deaf ears. The work of these new masters of modernist music only gained recognition abroad in countries more receptive to new ideas, such as the United States. Like many of their contemporary innovators in a Vienna caught between a faded past and an uncertain future, they found themselves prophets without honour in their own land.

**Klimt's beauty** *The paintings of the Austrian symbolist Gustav Klimt became synonymous with fin-de-siècle Vienna. His sensual depictions of the female body – as seen here in* Portrait of a Lady *(1918) but perhaps most famously expressed in* The Kiss *(1908) – imbued his works with a sense of uniquely decadent eroticism.*

## THE VIENNA SECESSION

The Vienna Secession was an art movement founded in 1897 by the painter Gustav Klimt (1862–1918). Its aim was to promote the work of young painters and others who were spurned by the city's conservative art establishment. Committed to regenerating art in Austria, the movement was instrumental in spreading Art Nouveau there. This international style became known in German-speaking countries as Jugendstil, from its main supporting journal, the Munich-based *Die Jugend* ('Youth'). The Secession, which brought together painters, sculptors, architects and designers, set up its own exhibition house near Karlsplatz in 1898, designed by the young architect Joseph Maria Olbrich. The distinctive and influential Secessionist style that emerged was typified by stylised floral ornamentation, sinuous curves and lines and a strong unity of images and typography in graphic works.

**Tragic death** *A poster by Egon Schiele, a protégé of Klimt, promoting the 49th exhibition of the Vienna Secession in March 1918. A brilliant painter, Schiele fell victim to the Spanish flu pandemic later that same year.*

143

# CHRONOLOGY

The timeline on the following pages outlines key discoveries and inventions of the late 19th and early 20th centuries. Selected historical landmarks are included to provide context for the scientific, technological and other innovations listed below them.

## 1893

- Ferdinand de Lesseps' Panama Canal Company goes bankrupt (1893)
- Women get the vote in New Zealand (1893)
- Outbreak of the Sino-Japanese War (1894)
- Opening of the Manchester Ship Canal (1894)
- Beginning of the Dreyfus affair in France (1894–99)

- Frenchman Léon Appert invents reinforced glass, which contains wire mesh

- American pharmacist Caleb Bradham invents Pepsi-Cola

- The world's first automobile, the 'Victoria', is built by German engineer Carl Benz

- German surgeon Carl Ludwig Schleich introduces local anaesthesia

- Chicago hosts the World's Fair of 1893

- The German firm of Hildebrand & Wolfmüller becomes the first company to mass-produce motorbikes

- The world's first motor race is held between Paris and Rouen

- British engineer Charles Parsons invents the marine turbine engine

## 1895

- Kiel Canal is opened, linking the North Sea with the Baltic (1895)
- In the Treaty of Addis Ababa, Italy renounces its claims on Ethiopia (1896)
- First Olympic Games of modern times are held in Athens (1896)
- Publication of The Jewish State by Theodor Herzl, the founder of Zionism (1896)

- German physicist Wilhelm Röntgen discovers X-rays

- First public showing of a motion picture by Auguste and Louis Lumière, using their 'Cinematograph'

- Russian physicist Konstantin Tsiolkovsky proposes that liquid-fuelled rockets could be used to propel vehicles into space

- American businessman King Camp Gillette unveils the safety razor with disposable double-edged blades

- British physicist J J Thomson identifies the electron, the subatomic particle that makes up electric current

- French physicist Henri Becquerel discovers background radiation in the Universe

- New Zealand-born British physicist Ernest Rutherford describes and names alpha and beta particles

- French cinematic pioneer George Méliès, known for his remarkable special effects, founds the world's first film studio

▶ The molecule heparin, discovered in 1916 by Jay MacLean, was eventually developed as an anticoagulant to prevent blood cells clotting

▼ Early windscreen wiper

▶ An experimental lamp used in the development of the 'Fleming Valve' or diode, a major step forward in radio technology

# 1897

- War breaks out between Greece and Ottoman Turkey over the sovereignty of Crete (1897)
- In the Spanish-American War, the victorious USA gains control of the Philippines and Cuba (1898)
- French and English colonial forces clash in the Sudan during the Fashoda Incident (1898)

- Dutch physician Christiaan Eijkman demonstrates a link between polished rice and the tropical disease beriberi

- The taximeter, invented by the German Wilhelm Bruhn, is installed in a cab for the first time

- Founding of the pioneering Liverpool School of Tropical Medicine

- Friedrich Löffler and Paul Frosch are the first to trace an animal disease to a viral infection

- Pierre and Marie Curie discover polonium and radium

- German chemist Eduard Buchner discovers zymase, proving that fermentation of sugar is caused by enzymes in yeast

- A vaccine against typhoid is introduced by the British bacteriologist Almroth Wright

- The first rigid-hulled airship is built by Austrian inventor David Schwartz, using an aluminium frame and sheeting

- Austrian chemist Adolph Spitteler develops galalith, an early plastic derived from the milk protein casein

▼ In a photo-reconstruction, a radio operator receives a distress call from the *Titanic*

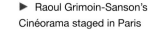

▶ Raoul Grimoin-Sanson's Cinéorama staged in Paris

# 1899

- The First International Peace Conference is convened in The Hague (1899)
- Outbreak of the Second Boer War in South Africa (1899–1902)
- The Boxer Uprising against foreign involvement erupts in China (1900)
- King Umberto I of Italy is assassinated by anarchists (1900)

- Glass windscreens make their debut on automobiles

- German chemist Felix Hoffmann rediscovers the anti-inflammatory and painkilling properties of acetylsalicylic acid; the remedy is marketed by the Bayer pharmaceutical company as Aspirin

- Gamma rays are discovered by Frenchman Paul Villard

- British inventor Frank Hornby patents a construction toy that achieves worldwide success as Meccano

- Austrian neurologist Sigmund Freud publishes his seminal work *The Interpretation of Dreams*

- Guglielmo Marconi establishes the first international radio link between England and France

- Canadian Reginald Fessenden invents the radio-frequency alternator, a prototype of the radio

- Clementines (a hybrid of bitter orange and mandarin) are cultivated for the first time in Algeria

- Maiden flight of the first Zeppelin airship, the *LZ-1*

- Raoul Grimoin-Sanson uses ten movie projectors – the Cinéorama – to show an audience a moving image in the round

- Max Planck publishes his *Quantum Theory*

◀ Édouard Branly's 'coherer', an early form of radio receiver

## 1901

### EVENTS

- Death of Queen Victoria; Edward VII ascends the British throne (1901)
- The first Nobel peace prize is awarded to Henri Dunant, founder of the Red Cross (1901)

### INVENTIONS

- Guglielmo Marconi transmits the first transatlantic radio message, from Poldhu in Cornwall

- Millar Reese Hutchinson invents the electrical hearing aid

- Dutch physicist Willem Einthoven develops the electrocardiograph

- French physiology professor Paul Richet is the first to identify allergies

- The speedometer is invented at around the same time by the British firm of Thorpe & Salter and the German engineer Otto Schulze from Strasbourg

- British engineer Frederick W Lanchester introduces disc brakes

## 1903

### EVENTS

- A treaty with newly independent Panama grants control of the canal zone to the USA (1903); the USA begins construction the following year
- Start of the Russo-Japanese War (1904–5)
- Great Britain and France forge closer military and diplomatic ties in the 'Entente Cordiale' (1904)

### INVENTIONS

- Nobel prize for physics is awarded to Marie and Pierre Curie and Henri Becquerel

- American Mary Anderson invents windscreen wipers for automobiles

- American toy designers Rose and Morris Mitchom and the German toy manufacturer Margarete Steiff produce and market the first teddy bears

- German chemists Joseph von Mering and Emil Fischer create Veronal, the world's first barbiturate

- Russian botanist Mikhail Tsvet devises a new method of analysis – chromatography – which finds widespread use in medical laboratories

- American aviation pioneers Orville and Wilbur Wright make the first successful controlled flight in a powered heavier-than-air machine

- 'The Landlord's Game', precursor of Monopoly, is created by Elizabeth Magie Philips in America

- Offset lithography, a new printing process, is introduced by American printer Ira Washington Rubel

▶ Model of a biplane built by French constructor Henri Farman

▼ Photographic plates of heartbeats made by the British physician Augustus Waller

▼ A hearing aid from 1929

# 1905

- A popular uprising in Russia is suppressed on Bloody Sunday (1905)
- Norway achieves independence from Sweden (1905)
- France and Germany vie for control of Morocco, which becomes a French protectorate (1906)

- Albert Einstein publishes his special theory of relativity, which leads scientists to reassess the concepts of space and time

- British physiologist William Baylis and Ernest Starling identify hormones

- The steam turbine-powered HMS *Dreadnought* is launched by Britain

- Norwegian Einar Berggraf invents the bathymetre, an early form of echo sounder and precursor of sonar

- Arsène d'Arsonval and George Bordas develop the technique of freeze-drying, which will find uses in food preservation and pharmaceuticals

- Enrico Forlanini tests his revolutionary new hydrofoil on Lake Maggiore

- French physicians Calmette and Guérin produce the BCG vaccine against tuberculosis

- The Simplon Tunnel linking France and Italy under the Alps is opened to rail traffic

# 1907

- Britain, France and Russia form the Triple Entente (1907)
- Revolt of the Young Turks in the Ottoman Empire (1908)
- Three-year-old Pu-Yi ascends the throne of China (1908)
- King Leopold of Belgium relinquishes power over the Congo (1908)

- Architects worldwide, foremost among them the Frenchman François Hennebique, begin to use reinforced concrete in their constructions

- French inventor Édouard Bélin develops an early version of the fax machine

- The Geiger counter for measuring radiation is introduced and named after its inventor, the German Hans Geiger

- The first Ford Model-T car is built in Detroit

- American inventor Elmer Sperry produces a gyroscope for ships

- Swiss engineer Jacques Edwin Brandenburger invents cellophane

◄ Albert Einstein explaining his theories

▼ A giant game of Monopoly

► Early animated figure by French cinematographer Émile Cohl

## 1909

- Serbia is forced to accept the annexation of Bosnia-Herzegovina by Austria-Hungary (1909)
- Death of Edward VII and accession of George V (1910)
- Union of South Africa formed as a British Dominion (1910)
- Annexation of Korea by Japan (1910)
- King Manuel II of Portugal is overthrown and a republic proclaimed (1910)

- Louis Blériot becomes the first person to cross the Channel by aeroplane

- First projection of a colour film, using the two-colour 'Kinemacolor' process devised by George Albert Smith

- 'SOS' is adopted as the international radio distress signal

- Belgian-born American chemist Leo Baekelend introduces the world's first successful synthetic plastic, Bakelite

- American Thomas Morgan highlights the role played by genes in the transmission of hereditary traits in the fruit fly

- Nitrogen fertiliser is synthesised by German chemist Fritz Haber

- The neon tube, invented by the Frenchman George Claude, is exhibited at the World's Fair in Brussels

- German engineer Hermann Föttinger devises a fluid coupling system, the prototype of the automatic gearbox

## 1911

- China becomes a republic; end of the Manchu dynasty (1911)
- The British ocean liner RMS *Titanic* hits an iceberg and sinks on her maiden voyage (1912)
- Balkan Wars (1912–13)
- The Triple Entente and the Triple Alliance (Germany, Austria-Hungary and Italy) engage in an intense arms race (1912–13)

- New Orleans emerges as the hub of jazz

- Polish-born American biochemist Casimir Funk identifies the first vitamin (thiamine, or vitamin $B_1$) in the husk of brown rice

- German meteorologist Alfred Wegener lays the foundations of his seminal work on continental drift and plate tectonics, *The Origin of Continents and Oceans*

- French physicist Charles Fabry reveals the existence of the ozone layer in the Earth's atmosphere

- Émile Bachelet successfully tests a magnetic levitation system, which later, from the 1960s onwards, forms the basis for maglev trains

- In Sheffield, stainless steel is developed by Harry Brearley

- Briton Arthur Wynne publishes, in the United States, the first newspaper crossword puzzle

▲ Early stainless steel butter knife

▲ A 1920s Art Deco powder compact made from Bakelite

▶ A magnified image of a genetically modified fruit fly

## 1914

- Opening of the Panama Canal (1914)
- Assassination of Archduke Franz Ferdinand of Austria in Sarajevo triggers the outbreak of the First World War (1914)
- First Battle of the Somme (1916)
- Bolshevik October Revolution overthrows Tsarist rule in Russia (1917)
- The Armistice brings the First World War to an end (1918)

- German aircraft maker Hugo Junkers builds the J-1, the world's first all-metal cantilever-wing monoplane

- American medical student Jay McLean discovers the anticoagulant molecule heparin

- The British unleash their new weapon, the tank, during the Battle of the Somme

- First use of shells containing mustard gas by German forces against Allied troops at the Second Battle of Ypres

- Dutch engineer Anthony Fokker devises interrupter gear, which allows a machine gun to fire through a spinning aeroplane propeller

- The world's first supermarket, brainchild of entrepreneur Clarence Saunders, opens in Memphis, Tennessee, under the name 'Piggly Wiggly'

## 1919

- The Treaty of Versailles is signed, imposing crippling reparations on Germany (1919); other post-war treaties dismantle the Ottoman and Austro-Hungarian empires and found the League of Nations
- Partition of Ireland sparks the Irish Civil War (1921–23)
- Adolf Hitler becomes leader of the German National Socialist Party (1922)
- Fascists seize power in Italy (1922)

- Deutsche Aero Lloyd, later to become Lufthansa, stages the first commercial flight in Europe, between Berlin and Weimar

- British pilots John Alcock and Arthur Whitten Brown make the first non-stop transatlantic flight

- British physicist Robert Watson-Watt patents his 'radiolocator', the prototype of his later more successful radar system

- Opening of the first motorway, the AVUS in Berlin

- Canadian John Larson develops the lie detector (polygraph)

- In France, Louis Rustin introduces a puncture repair kit for pneumatic bicycle tyres

- American Lee DeForest develops a system to record sound synchronised with pictures on a separate track, later leading to the first talking motion pictures

◀ Dropping a bomb from an aeroplane during the First World War

▼ Sigmund Freud (bottom row, left)

▼ Bicycle puncture repair kit

Rubber Solution
5gms e

## 1923

- Mustafa Kemal Atatürk becomes president of the new Turkish Republic (1923)
- Mussolini establishes a Fascist dictatorship in Italy (1925)
- Chiang Kai-shek assumes leadership of the Kuomintang, the Chinese Nationalist Party (1925)

- Clarence Birdseye founds the frozen food company that still bears his name

- Americans James Cummings and Earl McLeod build the first bulldozer

- The American company Kimberly-Clark introduces the disposable paper hankie

- German psychiatrist Hans Berger records the first electroencephalogram (EEG) of the human brain

## 1926

- Germany joins the League of Nations (1926)
- Emperor Hirohito ascends to the Chrysanthemum Throne of Japan (1926)
- Great Britain is paralysed by a General Strike (1926)

- Scottish inventor John Logie Baird creates the first recognisable television picture

- The era of spaceflight begins as American inventor Robert Hutchings Goddard completes the first successful launch of a liquid-fuel rocket

- German chemist Hermann Staudinger discovers that plastics are polymers, formed from long chains of small groups of atoms

- The age of the talking motion pictures dawns with *The Jazz Singer*, starring Al Jolson

- Aviator Charles Augustus Lindbergh makes the first non-stop solo flight across the Atlantic in his Ryan monoplane *Spirit of St Louis*

▶ An early electric iron

▲ Placing an order in Clarence Saunders' fully automated supermarket; although his self-service shops had gone down well with the public, this latest innovation from Saunders was a step too far

▶ *Die Großstadt* ('The Metropolis'), by Otto Dix, his iconic painting made in 1927 of the decadent jazz age in Germany's Weimar Republic

## 1928

- Stalin's first Five-Year Plan introduced in the Soviet Union (1928)
- Black Friday and the Wall Street Crash in New York mark the beginning of the Great Depression (1929)
- The Vatican City become an autonomous enclave (1929)
- Leon Trotsky expelled from the Soviet Union (1929)

---

- Chemical engineer Richard Drew invents Scotch tape

- British pharmaceutical company Smith & Nephew introduces a ready-to-use bandage called Elastoplast

- The radio beacon is introduced for aerial navigation

- German engineer Felix Wankel patents the rotary engine for automobiles

- First in-flight refuelling takes place between two aircraft

- Scottish bacteriologist Alexander Fleming discovers penicillin in mould

◀ Armchair designed by Josef Hoffmann, a founder-member of the Vienna Secession

◀ Paper hankies

## 1930

- Mohandas Gandhi launches his campaign of Civil Disobedience against the British colonial administration in India (1930)
- Japan invades the resource-rich northern Chinese province of Manchuria (1931)
- Franklin D Roosevelt begins his first term as US president (1932)

---

- American diver Otis Barton designs the bathysphere, a pressurised hollow steel ball for deep underwater exploration

- Discovery of the planet Pluto by astronomer Clyde W Tombaugh

- German scientists Ernst Ruska and Max Knoll build the first prototype of the electron microscope

- American astronomer Edwin Hubble formulates the 'Big Bang' theory of the origin of the universe

- German chemical concern IG Farben invents polyvinylchloride (PVC), a widely used new plastic

- British engineer Frank Whittle patents the jet engine

- Construction of the Empire State Building in New York

- The American DuPont chemical company introduces freon, the first commercial chlorofluorocarbon (CFC) for use in refrigerators

- Aviator Amelia Earhart becomes the first woman to fly solo across the Atlantic

▼ The Scheu House in Vienna, built by modernist architect Adolf Loos in 1912

# Index

Page numbers in *italics* refer to captions.

# Picture credits

ABBREVIATIONS: t = top, c = centre, b = bottom, l = left, r = right

**Front cover**: main image: The Art Archive/ Kharbine-Tapabor/Perrin Collection; inset: an early electric toaster: Cosmos/SSPL/Science Museum, London. **Spine**: nylon stockings: AKG-Images/ IC-Worldwide. **Back cover**: gas mask from the First World War, Dist. RMN/ É Cambier/© Musée de l'Armée, Paris.
**Page** 2, left to right, top row: Cosmos/SSPL/ Science Museum, London; AKG-Images/Sotheby's; Bridgeman Art Library/Offset Buch und Werbekunst, 1926 Joost Schmidt/Private collection; 2nd row: Cosmos/SSPL/Science Museum, London; Leemage/Selva; Dist. RMN/J Faujour, Jumo lamp/© National Museum of Modern Art, Paris, Collection Centre Pompidou/ DR; 3rd row: AKG-Images/Deutsche Lufthansa, 1931, Otto Arpke, Private collection, DR; Dist. RMN/É Cambier/© Musée de l'Armée, Paris; Corbis/Bettmann; bottom row: Cosmos/SPL/Daniel Sambraus; Stam Mart, Éditeur Thonet/Les Arts décoratifs, Musée des Arts décoratifs, Paris/DR, photo: Les Arts décoratifs/Jean Tholance; Dist. RMN/J-C Planchet/© National Museum of Modern Art, Paris, Collection Centre Pompidou.
**Pages** 4/5: Karlsplatz station, Vienna/Amana Productions/Photolibrary.com; 6tl: Leemage/ PrismaArchivo; 6tr: Cosmos/SPL; 6b: AKG-Images/Sotheby's; 7t: Cosmos/SSPL/Science Museum, London; 7cr: Corbis/Photocuisine/ Garcia; 7b: Corbis/Car Culture; 8tl: Cosmos/Geoff Tompkinson; 8bl: Bridgeman Art Library/Archives Charmet/DR; 8br: Cosmos/SSPL/Science Museum, London; 8/9t: Ernoult.com/A Ernoult; 9tr: Bridgeman Art Library/Offset Buch und Werbekunst, 1926 Joost Schmidt/Private collection/DR; 9b: Corbis/JazzSign/Lebrecht Music & Arts; 10t: Getty Images/Time & Life Pictures/Cornell Capa; 10bl: Corbis/Bettmann; 10/11: www.gregcirade.com; 11tl: Leemage/ Fototeca Affiche, 1956 par Sepo/DR; 11bl: Collection Christophel/Mickey © Walt Disney International; 11r: Bridgeman Art Library, Hangar, Orly, architect E Freyssinet/DR; 12tr: AKG-Images; 12l: AKG-Images; 12br: Bridgeman Art Library/ Archives Charmet/Conservatoire national des arts et métiers, Paris; 13tl: Dist. RMN/J Faujour, Jumo lamp © National Museum of Modern Art, Paris, Collection Centre Pompidou/DR; 13tr: TCD/ 'Flowers and Trees', 1932, Burt Gillett/© Walt Disney Production/DR; 13br: Corbis/LWA/Dann Tardif; 14l: Corbis/Robert Landau; 14c: Cosmos/ SSPL/H L Oakley; 14tr: Cosmos/SPL/Christian Darkin; 14/15b: Roger-Viollet; 15tl: Roger-Viollet; 15br: Dist. RMN/É. Cambier/© Musée de l'Armée, Paris; 16b: Bridgeman Art Library/Freud Museum, London; 16t : Rue des Archives/The Granger Collection NYC; 16r: Cosmos/SSPL/Science Museum, London; 17tl: Bridgeman Art Library/ Private Collection; 17tr: Corbis/Charles H Hewitt/ Hulton-Deutsch Collection; 17b: Corbis/Atlantide Phototravel/Massimo Borchi; 18/19: Roger-Viollet; 20tr: AKG-Images; 20l & br: Cosmos/SSPL/ Science Museum, London; 21t: Getty Images/Hulton Archive; 21b: Cosmos/SSPL/ Science Museum, London; 22l: AKG-Images/ Science Photo Library; 22r: Cosmos/SSPL/Science Museum, London; 23t: Leemage/Bianchetti; 23b: AKG-Images; 24t: AKG-Images/'Une seule manœuvre avec le SFER-20-Radiola', 1925, Leonetto Cappiello/© Adagp, Paris, 2010; 24b: Corbis/Bettmann; 25:AKG-Images/Ullstein Bild; 26t: Leemage/PrismaArchivo; 26b: AKG-Images; 27t: Cosmos/SSPL/NMEM Daily Herald Archive, London; 27r: Photononstop/Mauritius; 28t: Getty Images/Stone/Baerbel Schmidt; 28c: Leemage/ Heritage-Images/Land of Lost Content; 28cr: Leemage/Selva DR; 29t: Corbis/Photocuisine/ Garcia; 29b: Kharbine-Tapabor/Illustration Louis Poyet/Kharbine-Tapabor Collection; 30t: AKG-Images/Sotheby's; 30c: © Haribo; 30b & 31t: Cosmos/SSPL/Science Museum, London; 31cr: Cosmos/SPL/L Steimark/Custom Medical Stock; 31b: Cosmos/SPL; 32tr: Cosmos/SSPL/Science Museum, London; 32bl: Eyedea/Hoa-Qui/Age/ Georgie Holland; 33t: Corbis/Science Faction/ David Scharf; 33b: GettyImages/Stone/David Burder; 34tl: Cosmos/SPL/CNRI; 34b: Corbis/ Bojan Brecelj; 35t: Corbis/Car Culture; 35cr: Cosmos/SSPL/Science Museum, London; 35b: Corbis/Mika; 36t: Bridgeman Art Library/Archives Charmet, DR; 36b: Roger-Viollet; 37tr: Cosmos/ SPL/Oscar Burriel; 37cl: Corbis/Jerry Cooke; 38t: AKG-Images; 38b: Cosmos/SSPL/Science Museum, London; 39t: Ernoult.com/A Ernoult; 39b: Bridgeman Art Library/The Stapleton Collection/Private collection; 40t: Cosmos/SSPL/ Science Museum, London; 40b: Corbis/Bettmann; 41t: AKG-Images/Science Photo Library; 41b: Leemage/Imagebroker; 42tl: AKG-Images; 42r: Cosmos/SSPL/Science Museum, London; 43t: Bridgeman Art Library/The Stapleton Collection/ Private collection; 43b: Corbis/Bettmann; 44t: Ernoult.com/A Ernoult; 44cl: Leemage/Costa; 44bl: Corbis/Bettmann; 44/45c: Cosmos/SSPL/ Science Museum, London; 44/45b: Corbis/ Momatiuk-Eastcott; 45t: Ernoult.com/A Ernoult; 45br: Corbis/Gerolf Kalt; 46c: Cosmos/SPL/Mehau Kulyk; 46bl: Cosmos/Geoff Tompkinson; 47t: AFP/Getty Images/Ian Waldie; 47b: Bridgeman Art Library/Offset Buch und Werbekunst, 1926 Joost Schmidt/Private collection, DR; 48t: Cosmos/ SSPL/Science Museum, London; 48b: Getty Images/Chris Ware, Hulton Archive Collection; 49t: Rue des Archives/CCI; 49b: BSIP/Payet; 50t: Corbis/Randy Faris; 50c: Corbis/JazzSign/ Lebrecht Music & Arts; 51tr: AKG-Images; 51b: Getty Images/Charles Peterson, Hulton Archive Collection; 52t: © BPK, Berlin, Dist. RMN/Jochen Remmer, 'The Big City', 1927-8, Otto Dix, Staatsgalerie, Stuttgart/© Adgp, Paris, 2010; 52b: The Picture Desk/The Art Archive/Culver Pictures; 53t: Corbis/Bettmann; 53b: Corbis/Bettmann; 54tr: AKG-Images; 54b: Rue des Archives/BCA '2001: A Space Odyssey', Stanley Kubrick with Keir Dullea, 1968, Warner Bros; 55tl: AFP/Getty Images/Chung Sung-Jun; 55r: CNRS Photothèque/Cent/Lactamme/Jean-François Colonna; (56l: illustration, see below); 56r: AKG-Images; 57tr: www.gregcirade.com; 57bl: Cosmos/SPL/ Gustoimages; 58t: www.gregcirade.com; 58b : Corbis/NASA/Roger Ressmeyer; 59tl: Bridgeman Art Library/Archives Charmet/Private collection; 59tr: Look at Science/J Honvault; 59b: Cosmos/SPL/NASA; 60t: Corbis/Bettmann; 60c: AKG-Images/Science Photo Library; 61t: Getty Images/American Stock/Hulton Archive Collection; 61b : Leemage; 62t: Corbis/ Science Faction/Norbert Wu; 62b: Corbis/Bettmann; 63t: Leemage/Photo Josse, illustrator F Galais, 1918/Musée des Deux Guerres Mondiales, Paris, DR; 63b: Getty Images/Time & Life Pictures/Cornell Capa; 64: Leemage/Selva; 65tr: REA/Richard Damoret; 65tl: Leemage/Fototeca Affiche, 1956, par Sepo/ DR; 65b: Cosmos/SPL/NASA; 66tr: Getty Images/ AR Coster/Topical Press Agency, Hulton Archive Collection; 66bl: © CNAM/DAF/Cité de l'architecture et du patrimoine/Archives d'architecture du xxe siècle. Fonds Bétons armés Hennebique, Royal Liver Building, Liverpool, architect Thomas Aubrey/DR; 67t: Bridgeman Art Library, Hangar, Orly, architect E Freyssinet, DR; 67b: REA/Stéphane Audras; 68: AKG-Images/ Ullstein Bild; 69tl: REA/É de Malglaive, Unesco, Paris, 1994, architect Tadao Ando/DR; 69tr: Corbis/David/Lees, Exhibition Building, Turin, 1948-1950, architect Pier Luigi Nervi/DR; 69b: AKG-Images/Bildarchiv Monheim, Chapelle Notre-Dame-du-Haut, Ronchamp, 1955, architect Le Corbusier © F.L.C./Adagp, Paris, 2010; 70tr: Leemage/Selva; 70b: Leemage/Heritage Images/ Oxford Science Archive; 71tr: Leemage, dessin Émile Cohl, Private collection; 71cl: Collection Christophel/Felix the Cat, 1919, Otto Messmer Pat Sullivan/DR; 71b: Collection Christophel/Mickey © Walt Disney International; 72t: AKG-Images/ Walt Disney Productions, 'Snow White and the Seven Dwarves', 1937, Walt Disney © Walt Disney International; 72b: Corbis/Bettmann; 73t: AKG-Images Betty Boop © Paramount; 73b: Collection Christophel/'The King and the Mockingbird', 1980, Paul Grimault, Jacques Prévert © Les Films Paul Grimault; 74t: AKG-Images/Disney Enterprises/'Spirited Away', 2001, Hayao Miyazaki/© Walt Disney International; 74b : AKG-Images/Twentieth Century Fox/'The Simpsons', 2007, David Silverman/© Twentieth Century Fox; 75t: AKG-Images/Dreamworks/ Wallace and Gromit album, 2005, Nick Park et Steve Box/© Dreamworks; 75b: AKG-Images/Walt Disney Productions, 'Toy Story', 1995, John Lasseter/© Walt Disney International; 76/77t: AKG-Images; 76b: Leemage/MP; 77cr: Bridgeman Art Library/Look and Learn/Private collection; 77br: AKG-Images/Dodenhoff; 78t: AKG-Images; 78b: Corbis/Martyn Goddard; 79t: AFP/Marcel Mochet; 79b: Hemis/Franck Guiziou; 80: Kharbine-Tapabor Collection, 'L'Illustration', 09/02/1907; 81cr: AKG-Images/Science Photo Library; 81b: Bridgeman Art Library/Archives Charmet, Conservatoire national des arts et métiers, Paris; 82tr: Leemage/Bianchetti; 82tl: Getty Images/ Chaloner Woods, Hulton Archive Collection; 82b: Jupiter Images/Zedcor Wholly Owned © 2010; 83t: Corbis/Hulton-Deutsch Collection; 83b: Cosmos/SPL/Hank Morgan; 84t: TCD/'Flowers and Trees', 1932, Burt Gillett/© Walt Disney Production, DR; 84b: Cosmos/SSPL/NMEM Daily Herald Archive, London; 85t: Dist. RMN/J Faujour, Jumo lamp/© National Museum of Modern Art, Paris, Collection Centre Pompidou/DR; 85b: The Picture Desk/The Art Archive/Willard Culver /NGS Image Collection; 86: Bridgeman Art Library/ Private collection; 86/87: Corbis/Bettmann; 87tr: Getty Images/Photonica/Alison Shaw Veer; 87b: AKG-Images/IC-Worldwide; 88l: Getty Images/ NASA/Time & Life Pictures; 88tr: Rue des Archives/CCI; 89t: AFP/Anne-Christine Poujoulat; 89b: Corbis/EPA/Armando Babani; 90: Corbis/ Robert Landau; 91t: Roger-Viollet/Alinari; 91cl: Corbis/Hulton-Deutsch Collection; 92t: Corbis/ LWA/Dann Tardif; 92b: Corbis/Science Faction/ David Scharf; 93tl (x2) & c: AKG-Images/Science Photo Library; 93b Bridgeman Art Library/ Humboldt University, Berlin; 94l: Getty Images/The Image Bank/John Francis Bourke; 94r: © Images of the History of Medicine, United States, DR; 95tl: Cosmos/SPL/Pasieka; 95b: Getty Images/Visuals Unlimited, Inc/Dr Peter Artymiuk; 96t: Cosmos/ SSPL/H L Oakley; 96b: Leemage/Gusman; 97t: Corbis/Photocuisine/Maximilian Stock Ltd; 97b: AKG-Images/D Bellon; 98t: AKG-Images; 98b: Corbis/Bettmann; 99t: Leemage/Selva; 99b: Corbis/Bettmann; 100: Getty Images/Time & Life Pictures/Mansell; 101t: Bridgeman Art Library/ Peter Newark Military Pictures/Private collection; 101b: Rue des Archives/BCA; 102: Leemage/ Heritage Images; 101t: Roger-Viollet; 103cr: Rue des Archives/TAL; 103bl: Rue des Archives/TAL; 104t: Getty Images/Brooke/Hulton Archive; 104b: Leemage/Selva Fortune, 5/1933, Roger Duvoisin/ DR; 105t: Getty Images/Central Press/Hulton Archive; 105tr: AKG-Images/Deutsche Lufthansa, 1931, Otto Arpke, Private collection, DR; 106b: AKG-Images; 106t & 107cl: Cosmos/SPL/Christian Darkin; (107cr illustration, see below); 107b: Cosmos/SPL/Sinclair Stammers; 108: Corbis/ Atlantide Phototravel/Guido Cozzi; 108/109t: Cosmos/SPL/Gary Hincks; 109b: Corbis/Rykoff Collection; 110: REA/Laif/Marcus Hoenh; 111t: Cosmos/SSPL/Science Museum, London; 111b: Corbis/Bettmann; 112t: Rue des Archives/Mary Evans; 112b: Getty Images/Topical Press Agency/ Hulton Archive; 113: BSIP/Phototake/Kunkel; 114t: Leemage/Costa; 114b: Cosmos/SSPL/ NMEM Daily Herald Archive, London; 115t: Corbis/ Bettmann; 115bl: Cosmos/SSPL/Science Museum, London; 116t: Roger-Viollet; 116b: Corbis/Bettmann; 116/117b: AFP/E Ilan-Yediot; 118t: Dist. RMN/É Cambier/© Musée de l'Armée, Paris; 118b: AKG-Images/Ullstein Bild; 119t: Corbis/Hulton-Deutsch Collection; 119b: Roger-Viollet/Dod; 120cr: AKG-Images/Ullstein Bild; 120b: Bridgeman Art Library/Freud Museum, London; 121: Corbis/ Christies' Images, 'Psychoanalysis and Morphology Meet', 1939, Salvador Dalí/© Salvador Dalí, Fondation Gala-Salvador Dalí/Adagp, Paris, 2010; 122b: AKG-Images; 122/123: AKG-Images; 123t: AKG-Images/Ullstein Bild; 123b: AKG-Images/ Janos Kalmar; 124t: Roger-Viollet; 124b: Corbis/ Ron Chapple; 125t: Getty Images/ Herbert Gehr/ Hulton Archive; 125b: Cosmos/ SPL/ Daniel Sambraus; 126tr: Dist. RMN/ J-C Planchet/© National Museum of Modern Art, Paris, Collection Centre Pompidou; 126b: Leemage/Selva, Private collection; 127tr: Leemage/Selva; 127b: Bridgeman Art Library/Archives Charmet, Private collection; 128c: Cosmos/SSPL/Science Museum, London; 128bl: 'Chair without end', 1926-7, Stam Mart, Éditeur Thonet/Les Arts décoratifs, Musée des Arts décoratifs, Paris/DR, photo: Les Arts décoratifs/ Jean Tholance; 129t: AKG-Images/ Electa, illustrator Achille Mauzan/Private collection, DR; 129b: Cafetière Cona-Rex, 1950-60, Pyrex, Fabricant Kirby, Beard/Les Arts décoratifs, Musée des Arts décoratifs, Paris/DR, photo: Les Arts décoratifs/Jean Tholance; 130t: Rue des Archives/ The Granger Collection, NYC; 130b: Roger-Viollet; 131t: Kharbine-Tapabor/Collection S Kakou; 131b: Roger-Viollet; 132tl: Corbis/Bettmann; 132b: RMN/ Gérard Blot/Musée national de la coopération franco-américaine, Blérancourt; 132/133t: Getty Images/Aurora/Andrew McGarry; 133r: Getty Images/Photographer's Choice/Tony Hutching; 134b: Bridgeman Art Library/Private collection; 134/135: Corbis/Bettmann; 135t: Getty Images/ Francis Miller/Time & Life Pictures; 135b: Getty Images/Allan Grant/Time & Life Pictures; 136tl: Bridgeman Art Library/'Supermarket Shopper', 1970, Duane Hanson/Coll. Ludwig, Aix-la-Chapelle, Allemagne/© Adagp, Paris, 2010; 136tr: Dist. RMN/P Migeat, '99 Cents', 1999, Andréas Gursky/ © National Museum of Modern Art, Paris, Collection Centre Pompidou/© Courtesy: Monika Sprüth Galerie, Cologne/Adagp, Paris, 2010; 137t: Getty Images/The Image Bank/Terje Rakke; 137b: Corbis/Tony Latham; 138cr: Corbis/Charles H Hewitt/Hulton-Deutsch Collection; 138b: Cosmos/ SPL; 139: Corbis/Adrianna Williams; 140t: Leemage/Fototeca; 140b: Leemage/Selva; 141t: AKG-Images/Fabpics; 141cr: Getty Images/Imagno/ Hulton Archive, postcard, 1911; 142tl: Bridgeman Art Library/Private collection, Fauteuil, 1905, Josef Hoffmann/DR; 142b: Corbis/Atlantide Phototravel/ Massimo Borchi; 143hl: AKG-Images/Ehrt, Second Viennese School, 2004, Rainer Ehrt; 143r: AKG-Images/'Portrait of a Lady', 1917-18, Gustav Klimt/ Lentos Kunstmuseum, Linz; 143b: Bridgeman Art Library/Vienna Secession 1918, Egon Schiele/ Fitzwilliam Museum, University of Cambridge; 144/145: Corbis/JazzSign/Lebrecht Music & Arts; 146c: BSIP/Phototake/Kunkel; 146bl: Corbis/Mika; 146br & 147c: Cosmos/SSPL/Science Museum, London; 147bl: Corbis/Bettmann; 147br: Kharbine-Tapabor/Illustrator Louis Poyet/Kharbine-Tapabor Collection; 148/149t, 148bl, 148br: Cosmos/SSPL/ Science Museum, London; 149t: AKG-Images; 149bl: Leemage/Dessin Émile Cohl/Private collection; 149br: AFP/Getty Images/Ian Waldie; 150g: Bridgeman Art Library/Private collection; 150tr: Cosmos/SSPL/Science Museum, London; 150br: Corbis/Science Faction/David Scharf; 151l: AKG-Images; 151tr: AKG-Images/Ullstein Bild; 151br: Cosmos/SPL/Daniel Sambraus; 152l: Getty Images/Francis Miller/Time & Life Pictures; 152tr: Dist. RMN/J-C Planchet/© National Museum of Modern Art, Paris/Collection Centre Pompidou; 152br: © BPK, Berlin, Dist. RMN/Jochen Remmer, 'The Big City', 1927-8, Otto Dix, Staatsgalerie, Stuttgart /© Adgp, Paris, 2010 ; 153tl: Bridgeman Art Library/Private collection, Fauteuil, 1905, Josef Hoffmann/DR; 153bl: Corbis/Tony Latham; 153br: AKG-Images/Fabpics.

Illustrations on pages 56 (relativity) and 107 (Earth's continents today) by Grégoire Cirade.

**THE ADVENTURE OF DISCOVERIES AND INVENTIONS**
**Taking to the Air – 1900 to 1925**
Published in 2011 in the United Kingdom by Vivat Direct Limited
(t/a Reader's Digest), 157 Edgware Road, London W2 2HR

**Taking to the Air – 1900 to 1925** is owned and under licence from
The Reader's Digest Association, Inc. All rights reserved.

Copyright © 2010 The Reader's Digest Association, Inc.
Copyright © 2010 The Reader's Digest Association Far East Limited
Philippines Copyright © 2010 The Reader's Digest Association Far East Limited
Copyright © 2010 The Reader's Digest (Australia ) Pty Limited
Copyright © 2010 The Reader's Digest India Pvt Limited
Copyright © 2010 The Reader's Digest Asia Pvt Limited

Adapted from *Au Temps des Pionniers de l'Aviation*, part of a series entitled
L'ÉPOPÉE DES DÉCOUVERTES ET DES INVENTIONS, created in France by
BOOKMAKER and first published by Sélection du Reader's Digest, Paris, in 2010.

Reader's Digest is a trademark owned and under licence from The Reader's Digest
Association, Inc. and is registered with the United States Patent and Trademark
Office and in other countries throughout the world. All rights reserved.

All rights reserved. No part of this book may be reproduced, stored in a retrieval
system, or transmitted in any form or by any means, electronic, electrostatic,
magnetic tape, mechanical, photocopying, recording or otherwise, without
permission in writing from the publishers.

**Translated from French by** Peter Lewis

**PROJECT TEAM**
**Series editor** Christine Noble
**Art editor** Julie Bennett
**Designer** Martin Bennett
**Consultant** Ruth Binney
**Proofreader** Ron Pankhurst
**Indexer** Marie Lorimer

**Colour origination** FMG
**Printed and bound** in China

**VIVAT DIRECT**
**Editorial director** Julian Browne
**Art director** Anne-Marie Bulat
**Managing editor** Nina Hathway
**Picture resource manager** Sarah Stewart-Richardson
**Technical account manager** Dean Russell
**Product production manager** Claudette Bramble
**Production controller** Sandra Fuller

We are committed both to the quality of our products and the service we provide to our
customers. We value your comments, so please feel free to contact us on 0871 3511000
or via our website at **www.readersdigest.co.uk**

If you have any comments or suggestions about the content of our books, you can
email us at **gbeditorial@readersdigest.co.uk**

CONCEPT CODE: FR0104/IC/S
BOOK CODE: 642-009 UP0000-1
ISBN: 978-0-276-44521-7